THE LONG SPRING

THE LONG SPRING

Tracking the Arrival of Spring Through Europe

Laurence Rose

Illustrations by Richard Allen

B L O O M S B U R Y

LONDON • OXFORD • NEW YORK • NEW DELHI • SYDNEY

For Margaret Rose
in memory of Bernard Rose
(1927–2008)

BLOOMSBURY WILDLIFE
Bloomsbury Publishing Plc
50 Bedford Square, London, WC1B 3DP, UK

BLOOMSBURY, BLOOMSBURY WILDLIFE and the Diana logo are trademarks of Bloomsbury Publishing Plc

First published in Great Britain 2018

A catalogue record for this book is available from the British Library.

Library of Congress Cataloguing-in-Publication data has been applied for.

ISBN: HB: 978-1-4729-3667-7
ePub: 978-1-4411-3670-7
ePDF: 978-1-4411-3669-1

2 4 6 8 10 9 7 5 3 1

Illustrations by Richard Allen

Typeset in Bembo Std by Deanta Global Publishing Services, Chennai, India
Printed and bound in Great Britain by CPI Group (UK) Ltd, Croydon CR0 4YY

Contents

Spain

1 February 2016, Ceuta. 35° 54' N. Looking south from Monte del Renegado I see all of Africa like a map unscrolled stretching away into the glare.

What my eyes see is bounded by the Rif: Mount Boanan and its neighbours beyond Tétouan 30 miles to the south. In my mind, the view stretches past the limestone peaks to the burning sands of the Sahara, and beyond that, to the thin sanctuary of the Sahel. Unseen, but more than imaginary, it is something sensed. The Congo Basin sweats dark and fragrant before the great watershed at the threshold of the continent's south. Farther still, beyond the sea fog where scorched Namib and icy Atlantic tryst; I see beyond the Okavango swamplands and southern savannas, all the way to the meadows of the Cape.

The dust and dews of Africa are bound into the flesh of 2 billion birds, now beginning their journeys to the lands to the north. They may be European by breeding, but they are African in make-up. Every sinew and muscle that propels them here, all the fat gained as fuel for the journey and most of the feathers of their aero-architecture, were replenished in Africa. The birds that end their days in some English field or Finnish *aapa* will bring a morsel of the rainforest to the northern soils, and their progeny will return it some future autumn. I look to the south as if to see into their eyes. I sense that a few of them may already be here … perhaps swallows, thinly spread along these forested bluffs or strafing sheep flocks in Morocco, just beyond sight. The rest have yet to reach the valleys of the Atlas, most still to confront the thirsting wastes of the Sahara. In their hundreds of millions, they bedeck the acacias of Niger and Mali. Still more are irrupting out of the reed beds of the great deltas or plying northwards through the dripping canopies of Gabon and Cameroon.

Here, at the northern tip of the continent, the chill is lifting, and with it, the tang of herbs and goat shit, acrid and sweet and nostril-cloying, and dragonflies have appeared to ride the waves of rising air. I turn to face north, my gaze panning past the fog-white buildings of Ceuta and its bastion hillock Monte Hacho. With the sun behind me, I see across the Strait of Gibraltar all the cooler colours of Europe. This is instantly the difference between the two continents. Africa: white and platinum and etiolated purple; Europe: reflecting Africa's sun in uncut sapphire and emerald. A strait separates them that today is the colour of denim, neither calm enough to return the sun's glare in full nor rough enough to drag the blue of the ionosphere to its depths.

There are no early swallows today. But as the air warms, the handful of dragonflies becomes tens, then hundreds – migrant hawkers, local in origin, but preparing to cross the Strait into Europe when the time comes, when wind direction and warmth allow. The patterns they describe around us are a kind of needlecraft, tatted among the white asphodels as if insect and flower were collaborating on some invisible lace frivolity.

We came here, Jane, José and I, along Arroyo de Calamocarro, a stony stream bed tucked into the cleavages of the lower slopes, through a forest of oleander, wild olive and cork oak. José, a native Ceutí, has been keen to show off the local specialities, including African birds found nowhere else in Spain. He picked up a call from within a stand of giant reeds, a three-syllable trill, to my ears not unlike the chirrup of a budgerigar but with a mellow, thrush-like timbre.

'Listen! Bulbul!' The word is Persian, appropriated from an Arabic name for the unrelated nightingale, and adopted by the British and Spanish. The 130 or so bulbul species are found across Africa, the Middle East and Asia.

This one, the garden bulbul *Pycnonotus barbatus*, called from behind us, across the narrow valley of the arroyo. We turned, but it proved elusive, hidden in a thicket of wild olive.

The empty stream made no sound but all around was a watery tinkling of serin song that seemed to occupy the vacant sonic niche. Several males were singing from their high perches into the natural amphitheatre of the lower Calamocarro. As if compelled by an uncontrollable ecstasy, their usual rapid, tinkling bell notes were hissed out at high pitch and speed, like a leak in a pressure valve. Focussing my binoculars on one, I saw what looked like a yellow plum, puffed to bursting point, twisting from side to side in a short arc to spread its jingles across the valley.

'*Herrerillo!*' announced José at the sight of a pair of birds, smaller even than the serins, crossing over our heads from a eucalyptus to a small trackside shrub. I was keen to see this bird because since my last visit to north-west Africa the genetics of the humble blue tit had been studied in detail. As long ago as 1841, Charles Lucien Jules Laurent Bonaparte (1803–1857), Prince of Canino and Musignano and a nephew of Napoleon, described the distinct race of blue tit that inhabited the farther shores of the Mediterranean. Bonaparte had fled to Rome as a child under the protection of Pope Pius VII, and there developed a passion for ornithology. After the fall of his uncle, he eventually settled in Philadelphia, where he worked on his four-volume book *American Ornithology*, in which many species are described scientifically for the first time. Among them is *Chroicocephalus philadelphia*, still known today as Bonaparte's gull.

Bonaparte noticed what I was hoping to see – a distinctly different-looking bird compared to the blue tit from home. As they moved in and out of view, as busy and curious as you'd expect, I saw that their wings and back were a uniform ultramarine blue, and they sported dark caps that appeared black in the morning light. In size, behaviour and overall pattern, they were indistinguishable from the blue tit I know.

But the colouring-in of the pattern pointed to a variety that has been separated from its European counterparts for a long time. In 2005, a group of American scientists led by Frank Gill – coincidentally working at a university in Bonaparte's Philadelphia – analysed the birds' DNA to work out exactly how the tit family evolved and what their taxonomic relationships were. The results were unexpected. They had all shared the scientific name *Parus* since Linnaeus coined it in 1758. Now it seems they are not so closely related after all, and the familiar set of garden species have been given new generic names. *Parus* is retained by the great tit, *Parus major*, while the little coal tit is now called *Periparus*. The blue tit is *Cyanistes*, from the Classical Greek *kuanos*, meaning dark blue. Then, in 2008, a team of molecular biologists from Germany and ornithologists from Tenerife studied *Cyanistes* in more detail, and found that the African and Canarian forms were genetically distinct enough from all the other blue tits in Europe to consider as a whole separate species: *Cyanistes teneriffae*, the African blue tit.

'One continent, two flavours,' says José, as we stand on the slightly domed roof of the *fuerte,* the semi-sunken fort, like a large pillbox, which marks the summit of Monte del Renegado. The death mask of Jebel Musa is not much more than a raven's call away to the west. Here they call the Jebel La Mujer Muerta (the dead woman). The double summit of her head and her breast; her closed eyes; the tear on her bony cheek; her perfectly proportioned nose, lips and chin; the dip of her neck; the clasped hands on her chest, raised at the top of her last breath: these are visible to the inhabitants of this enclave of Spain, but not to the Moroccans on whose soil she lies.

'Raptor!' Jane calls from the ring of flat ground at the base of the fort.

'*Ratonero moro*!' José identifies the long-legged buzzard. He had been hoping we'd see one, a species he has studied carefully over the years, mapping territorial boundaries and nest-sites. It is subtly more elegant than the common buzzard, which, José explains when Jane asks, is uncommon here. Long- and straight-winged, it flies away from us towards the south, and gliding below the plane of our elevated position, its wings are silvered by the sun's glint.

I want to know more about why, last night, when we met for the first time, José had turned up at our hotel reception wearing what appeared to be a full-length embroidered dress in turquoise silk, pale foundation make-up that had lost the battle with his five-o'clock shadow, and a disc of scarlet rouge the size of a euro coin on each cheek. Jane and I had arrived there two hours earlier, and could soon tell something was going on. In a brief twilight, from our hotel balcony we could see east to Monte Hacho, the low fortified hill that is the province's seaward-most extremity. Straight ahead, 9 miles due north, Gibraltar wallowed pink-jowled in the evening light. To our left, the view was blocked by an apartment building whose walls echoed the wolf-whistles and glissandi of twenty or so spotless starlings that gathered on rooftop aerials like artificial leaves, coal-black in the gloaming light. The buildings seemed to funnel and refract sound from the streets below. We became aware of a larger gathering of souls six storeys below us, out of sight, a shimmering sound like waves on a pebbled shore, building into a murmuration of hailings and conversation, and a long, slow crescendo as if a thudding bass drum were to sound the climax at any moment.

Which it did. A cheer went up, cut short by another drumbeat, repeated to a steady pulse and interspersed on the half-beat by cymbals. A few bars of this simple rhythm heralded an announcement, unintelligible at our elevation, declaimed in the style of a music-hall chairman. It was met with cheers, laughter and catcalls, which were in turn interrupted by another rhythm on the percussion: a loud, coarse paso doble. A choir launched into song. At first, it sounded rough and dissonant, but as my ear grew accustomed to the urban acoustic and filtered out the distortions and echoes, I became aware of a fine tenor lead voice and accomplished close harmonies. Each song was greeted with cheering and laughter, and eventually, the show culminated in a rendition, by everyone present, of Tom Jones's 'Delilah', a kind of fandango after all.

I had a phone call to make. José Navarrete Pérez (the Ceuta representative of the Spanish conservation group SEO/BirdLife) and I had agreed by email that I would call him when we arrived to discuss plans for today's trip.

'We're in the Hotel Ulises.'

'I'm nearby, I'll see you in reception in one minute.'

When we arrived at the ground floor, all we could see were some recent Japanese arrivals and a middle-aged couple dressed in silk djellabas. I introduced Jane, and José introduced his partner, Antónia, who wore a peach-coloured djellaba and a tiara made of gold chains. José, a tall, slim man in his fifties, aquiline and with thin, short-cropped greying hair, apologised for his croaky voice: he had been singing all afternoon. Antónia explained that it was the start of carnival week, and apologised that they had to get back to their group for the next set. José told us where he would pick us up, and that we were to look out for a green Hyundai at nine in the morning – this morning.

We walk back to where José parked the car, and he explains that birds and *carnaval* are his two greatest passions.

'Under Franco, *carnaval* was banned,' he explains, 'so as children we sang only at Christmas. Cádiz was the only place in Spain where a bit of *canto carnavalesco* went on, behind closed doors. I loved singing at Christmas, and when *carnaval* was allowed again, here in Ceuta we adopted the Cádiz styles, and I was hooked. My group sings *chirigotas*: they are street songs with different lyrics every year – humorous, satirical. Yesterday was a try-out of some of our material, but the big event comes in a few days' time when six groups from here and another four from across the Strait compete in the Concurso de Agrupaciones Carnavalescas.'

'How long have you been interested in birds?' I ask.

'Always. I was a *pajarero*.'

I had come across the word before, as slang for 'birdwatcher'. Once, helping with a survey of flamingos in Doñana, I was accused of being a true *pajarero* when I broke off to try to identify a small warbler hidden in the scrub. But the real meaning of the term is 'bird-catcher'.

'You know, with nets,' said José, eliminating any doubt.

'Did you catch them to sell, as a business?'

'No, for food. Finches mainly, in the bushes around here. Then when I was eighteen, I joined the army, and it was no longer possible. So I started to watch and study them and found I preferred that. I spent twenty-nine years in the army, but now I spend all my time studying birds and rehearsing.'

2 February, Strait of Gibraltar. 36° 02' N. The sea has a gently rippled surface, like the silky phyllite stones we saw yesterday resting on the dry bed of the Arroyo de Calamocarro. I am in the lee of the port stack of the *Passió per Formentera*, a ferry of the Baleària line, looking back at Ceuta and La Mujer Muerta. I am looking west, and from here there is symmetry in the sweep of the mountains as they curve around from the south and dip below the surface of the Strait, the Rif of Africa reappearing on the European side as the Baetic ranges. Together they form what geologists call the Gibraltar Arc. Geologists sense time differently, and think of the open narrows like eyelids, halfway into a geo-temporal blink that will see the Strait close again a short time hence. They are closing now, at a rate of half an inch a year, and once fully shut, a mere thousand years later the Mediterranean Sea will have evaporated away to nothing but a salt and gypsum enamel. It happened 5.9 million years ago, and half a million years later, the Atlantic was allowed back in, refilling the Med in under two years.

I have plied these waters before, in July 1978. That year I left home with a rucksack and mild toothache, and a clutch of tickets for a succession of slow trains from London Victoria to Algeciras. I was intent on a few weeks in Morocco before returning for my second year at university. When I reached the Mediterranean more than two days later, I was nursing a swollen face and could think of little but my own sorrow. I remember nothing of the outward ferry across the Strait, and once in Tangiers sought nothing beyond getting through the night and to a dentist the next day.

By the following morning, I had been fleeced by my *pension* landlady and robbed of my camera in the street, and could barely stand upright as my infected tooth caused my head to spin. I hailed a taxi and returned to the ferry port. In my burgeoning self-pity, I decided to trust no one this side of Madrid, which is where I would go to find a dentist.

The ferry looked dirty and crowded, so despite my poverty, I bought a first-class ticket and made my way to a small upper deck. An English-looking man in his early middle age, casually dressed with casually combed, greying blonde hair, was sitting reading a small, navy-bound volume of close-typed French prose.

'Is this first class?' I asked.

'This boat doesn't have a first class,' he said.

'Bastards!'

'That doesn't stop them selling you a first-class ticket. If you ask nicely.' The Englishman pointedly avoided averting his eyes from his book, affecting a dismissive and offhand air, as if affronted that I had interrupted his reading. I moved towards the side of the deck, and the ferry got under way. As soon as we left the harbour and were in open water, a pod of five dolphins came alongside. The sunlight on their wet, flawless skins gleamed like satin each time they broke the surface. Dolphins are unfailing lifters of spirits, and I was in need of their cheer. The Englishman joined me at the rail, and we shared the duty of spotting the animals each time they reappeared. He had mellowed at the sight of them, and we struck up a conversation, discovering a shared interest in wildlife.

'I've been exploring Patagonia,' he said. 'I spent months there a few years ago. Beautiful, in its way, still wild and remote. I found a cave full of fossilised ground sloth shit.'

'So … you'll be Bruce Chatwin, then?'

Only weeks before, I had read *In Patagonia*. In number ninety-three of the book's ninety-seven chapters, Chatwin has crossed the Strait of Magellan to visit Punta Arenas in Chile, where his grandmother's cousin, Charles Milward, lived in the first decades of the 1900s. Chatwin arrived in Puerto Consuelo and walked four miles to a cave where Milward had helped a German gold prospector, Albert Konrad, dynamite the cave to reveal the remains of numerous giant ground sloths, or mylodons: fur, bones and semi-fossilised dung. The mylodon was a huge, bear-like herbivore that weighed about 2.5 tons. It probably went extinct over 5,000 years ago, but the cold, constant conditions in the cave kept the remains looking fresh.

As we watched the dolphins, leaning on the rail, Chatwin clutched his volume of Jean Racine in his right hand; I clutched my grubby copy of Heinzel, Fitter and Parslow's *Birds of Britain and Europe with North Africa and the Middle East*. He asked to see it and started to flick through pages black with sweat and the dust of the Negev, the grimy legacy of the previous summer's birding.

'I have a house in Spain, and there are loads of birds there I don't recognise.' Now and again he would point to a picture in the book. 'I think I might have seen that one – is that likely?'

He handed the book back and wondered if he would be able to buy a copy in Madrid. 'I'm not sure,' I said, 'but why don't you keep that one.'

'Or, you could come and spend a few days at the house and identify them all for me.'

The world in 1978 was ignorant of the disease that would kill Chatwin a decade later; only when I read his obituaries did I learn of his reputation for seducing young men. I took his suggestion of a visit at face value, although I was too straight and too innocent to know the signs of a proposition. In any case, toothache prevailed, and I told him, untruthfully, that I had a ticket to London that I had to honour. He took my book but insisted on paying for it, fishing 200 pesetas from the pocket of his shorts.

About halfway across the Strait, the sea now has a swell to it; the gentle rippling of the inshore waters has developed into a succession of low unbroken waves. I notice a line of dark-brown birds flying parallel to us. They are effortless in their stiff-winged air-skiing, inches above the waves. There are thirteen, evenly spaced in a line, letting the friction between water and air generate the effort they need to propel them forward. The line they follow is an undulating one, like a slow melody made visible, and I realise they are working layers of air in the same way a cyclist might requisition the energy from a downhill stretch to supplement the effort needed for the next rise. They flap their wings briefly every few seconds and at the same point in their trajectory, like a line of cars driving over a hump in the road.

There is only one group of birds so at ease in the space between the waves and the wind. They are known to ornithologists as Procellariiformes, from the Latin *procella*, a violent wind, and all members of this diverse order think nothing of spending months at a time far from the smell of land, taming the wildest storms. The smallest members are the storm petrels; the largest, and the greatest of all ocean wanderers are the albatrosses. In the middle ranks are the shearwaters, which by day inhabit a slab of air a few feet thick and a slab of water a foot or two below it. At night, for the few weeks of their breeding time, they return to burrows dug into islets far from the mainland. These thirteen all have upperparts the colour of drinking chocolate powder but are not all alike. Four of them are almost uniformly brown all over while the others have pale undersides, and I notice that even the dark varieties have a small, ill-defined off-white patch in the middle of their bellies. The palest ones have at least a smoky smudge under their tails and all look less than pristine below – the identification is confirmed.

These are the birds I have been most hoping to see in these waters, the critically endangered Balearic shearwater. It is a species with a tiny breeding range and a small population, numbering around 3,000 pairs and undergoing an extremely rapid decline. At their breeding colonies on islets off the main Balearic Islands, they are eaten by introduced mammals. At sea, they are killed as – to use a rather lame euphemism – 'fisheries bycatch'.

When a population is as vulnerable – as small and highly concentrated – as this, the risk of extinction due to sheer bad luck is high. The relatively new science of population viability analysis makes it possible to estimate how long we can expect it to survive as a species, given the rate of decline and modelling a variety of random environmental and demographic catastrophes. Such models estimate a mean extinction time of sixty years.

3 February, El Rocío, Doñana, Andalucía. 37° 07' N. A zest of moon hangs low over the tamarisks, casting enough light to silhouette them and to create a false dawn in the eastern sky.

I am standing at the edge of La Madre de las Marismas (The Mother of the Marshes), a shallow lagoon three-quarters of a mile across. I have not had to walk far in the darkness to get here: behind me is El Rocío, a white-walled village built on the prehistoric sands of this vast wetland, whose far edge is beyond the rim of the earth. It is seven o'clock, and sunrise is due an hour and a half from now. Astronomers have calculated that this is the moment the morning twilight begins, when the sun is eighteen degrees below the horizon; but I failed to reckon with the moon, which rose three hours ago. Milky, moon-reflected sunlight has been shed across these flat acres since long before I emerged from my bed in a small hotel a few yards away.

I had hoped to begin in darkness, to witness a sound-world slowly making way for a sight-world, the defining transformation of the spring dawn. But already greylags are calling from the middle of the marsh and a robin in the village is singing to the sodium lamps. Every few seconds something calls: moorhen, lapwing, dog, horse; a tawny owl on the far side of the lagoon. After ten minutes I hear the gently thudding hooves of a walking mare. A teal gives four amphibian chirrups. There is a strange sound, like a thin stream of water from a pipe dropping onto mud. A moorhen gives a high-pitched croak; a large insect flies close, buzzing deeply on two pitches. The mare tears at the grass with a gentle rhythm.

Another ten minutes: a harsh heron call and a lapwing skirl, both from overhead and coming close, then an immediate flurry of sharp, metallic, staccato scolding from black-winged stilts and coots. Then, a sudden quiet, save only for the strange mud-puddling sound. Soon there is enough light to reveal its source. Two teal are dabbling in the mud, which is covered by a mere film of water. They are working side by side, advancing towards me a foot or so every minute. They have emerged from deep in the tamarisks, and have left a record of their journey in the mud, like a tyre print.

The mare is moving again, then the hoof steps stop and the sound of her urine blends with the teals' mudwork. Now everything from the day-world of the marsh is calling: egrets, coots and glossy ibises, grunting as they emerge from their roost in the tamarisks, and a pumped-out note from a moorhen. A mares' chorus starts from the drier ground far to my left, and in response, a herd appears from the marsh to my right, vocally silent, but there is a loud swishing and paddling, and the clop of hooves through the mud. The greylags take off; their honking circles the marsh and is suddenly loud and overhead. In the foreground, a coot's short, air-piercing squeak repeats at an insistent pulse. A blackbird's rapid-fire wake-up call. A sudden bursting declamation from a Cetti's warbler. A cock crows at last, a long way off to the west. The teals' mud-larking continues.

I return to the hotel for breakfast, and then Jane and I walk to the far corner of the lagoon, keeping El Rocío to our right and the marshes to our left, before bearing left to leave the village behind and skirt the western shore. Diagonally opposite this morning's vantage point is Las Rocinas, a slow, reedy stream that flows from the west into the marshes. From here, looking north-east, El Rocío forms the backdrop to the lagoon. The church, the Ermita de la Virgen del Rocío, commands the skyline, her *espadaña* (or bell-gable) hoisted like a foresail.

Swallows have arrived. There are a dozen or so, sleekly dipping and hawking, scribing their blue signatures in the air. I follow one through binoculars for a while. His long tail-streamers are the insignia of his sex, and embellish his every figure; he strikes angles against the sun that glint blue-silver. Each split-second bank and stall in his flight path is a blaze on the trail of his invisible prey, and also a signal of athletic prowess. As my swallow loses himself halfway across the lagoon, I lower my bins and take in the scene again, imagining the constantly changing perspectives from which so blithe a traveller must view it. His close focus and high-definition imaging of minute quarry, his instinct for traffic control in shared airspace, the way he partitions the landscape for hunting or sleep … the knowledge he holds of what lies over the northern horizon and of how to traverse the lands to the south. I cannot know how much more of Europe must pass beneath his wings before he is home, how many more times our paths might theoretically cross. He is likely to stay here a while, though. Doñana is where spring arrives first in Europe; a swallow may leave Africa knowing that a mild and insect-rich welcome awaits in these *marismas.*

The shallow water has attracted thousands of pintail and teal from the north, and they show no sign of impatience for the return of spring. Groups of spoonbills, local breeders mostly, are sweeping for crustaceans through a wide arc, the foraging strategy of metal detectorists. Greater flamingos, whose daily, seasonal, annual and multi-annual patrols of the Mediterranean conform to no pattern discernible to us, but which appear none the less to serve some cause and purpose. Black-winged stilts, impossibly leggy, whose black and white plumage and pink legs and bill explain their Spanish name *cigüeñuela*: storklet. They are opportunist wanderers. A wet Doñana is paradisiacal, assuaging each species' every need: it is a bacchanal of feasting and fecundity. In a poor year, the kind of dry year in three that Doñana has always seen, the storklets will leave and wander until they find the conditions they require. In some years, a pair or two arrive in England and breed.* Today, 3 February, is when we are supposed to judge whether it will be a good year.

* In 2017, an unprecedented 13 black-winged stilts fledged from four pairs in Kent, Cambridgeshire and Norfolk.

Por san Blas,
la cigüeña verás;
si no la vieres,
año de nieves.

Come San Blas Day,
storks on their way;
if they don't show,
winter of snow.*

For a snowy winter in the *sierras* heralds a good year in the lowlands. While it may hold up the return of the storks from Africa, it means that water will trickle down onto the plains and wetlands watering crops and replenishing fisheries. This San Blas Day the storks are with us, and the water in La Madre de las Marismas and Las Rocinas is low. It has been a dry winter.

For the afternoon of San Blas, we have come to a small estate across the marshes from El Rocío, called Dehesa de Abajo. It is on a tongue of higher ground that points south into the vast levels that were formed millennia ago by the slow manoeuvring of silt and side-winding estuaries embattled with sand and dune-building wind. Immediately below our lookout, the natural wetlands have been fixed in position and hemmed in by land where rice and cotton grows, land redrawn into rectangles decades ago by the Food and Agriculture Organization of the United Nations. Half a century before that, two British residents, Abel Chapman and Walter Buck, wrote *Wild Spain* (1893) and *Unexplored Spain* (1910). Looking back upon a life in the field, in 1927 Chapman wrote:

> From time immemorial the Coto Doñana had been recognised as one of Spain's most famous hunting-grounds,

* I have taken a few liberties for the sake of the rhyme. Literally, the refrain goes: 'By San Blas Day, you will see the stork; if you do not see it, snowy year.'

> during centuries a favourite resort of Spanish kings. In physical formation this semi-isolated region is almost unique. It forms, in a sense, the Delta of the Guadalquivir and occupies the whole space between that great river and the Atlantic outside – in length something like forty miles with a varying breadth that is more or less indefinite – and practically cut off from the mainland by the 'marisma'. Its main constituent is wind-blown sand, presumably overlying the deposits of alluvial soil brought down during ages by the Guadalquivir. At its seaward extremity, towards the embouchure of that river, the Coto is overgrown with pine-forests, all embedded in luxuriant undergrowth. The central region is a strange blend of silent Saharan wastes and mountainous sand-ridges, between which are interposed long straggling belts of stone-pines and scrub-jungle – weird landscapes of indescribable beauty.[1]

We have come to this outcrop because white storks nest here, in the greatest abundance of any place in Spain. In Dehesa de Abajo, around 400 pairs share the *acebuche* (wild olive) and stone pine trees with a colony of black kites. This San Blas Day, most of the huge and venerable stick platforms are already occupied. That is no surprise: 3 February may never have been a typical first date for storks, which can often be seen arriving back before Christmas; and with a changing climate, the old refrains are ever less appropriate. But the biggest change in stork behaviour has come with the surge in domestic waste, and a growing proportion of storks no longer leave Spain at all. Of those that do, some carry names and satellite tracking devices and send back news of their whereabouts in something close to real time.

'Picopelucho' is a young stork who was tagged in his nest here on 20 June last year. A month later he left the area and crossed the Strait of Gibraltar on 22 July with thousands of others. He stayed in Morocco for a month, then spent a week crossing the Sahara, arriving in the Sahel on 7 September. He spent the winter 1,400 miles from here, moving around the semi-desert and savanna between Mali and Mauritania. This research demonstrates the tension between instinct and opportunity. The majority of tagged adults stay in Spain, having learnt that rice fields and rubbish dumps offer at least as good pickings as the tropical African savanna. The majority of young birds migrate, drawn by instinct and knowing no better.

At El Rocío, Perez's frogs start the evening chorus, a long, tenuto croak, dozens of overlapping calls laying down a constant, barely pitched drone. Another species that I cannot identify superimposes an upwards-inflected spiccato, repeated countless times, then stops without warning. Glossy ibises assemble at the base of the tamarisk thicket before choosing a branch for the night. Small murmurations of spotless starlings, a hundred or so at a time. A cricket calls from an exposed patch of grass in the marshes like an unanswered telephone; another is in a nearby tree at the edge of the village, calling back in loud, twenty-second bursts of trilling; their combined vibrations cause the air to shudder. Lapwings skirl, snipe grate, black-winged stilts bicker with loud, sharp, staccato complaints. The short, emphatic song of the Cetti's warbler, which always sounds unfinished, is at its amateur best at dusk – the bastard nightingale, to give it its Spanish name.

4 February, El Rocío. At the edge of La Madre de las Marismas, a rapid movement among the green-shooting flag irises catches my eye. A male bluethroat, like a tail-cocked robin, wearing a black breastplate starred with a few new blue feathers, is foraging at the edge of a micro-pond in a horse's hoof-print. He watches me for a second, then rises deliberately and mechanically by unfolding his surprisingly long legs. And then he just vanishes.

I head north from the shore and into the village. I pass the open west door of the church-hermitage and glance in. There she is: *La Virgen del Rocío* (The Virgin of the Dew), at the far end behind her altar. Petite, pretty, lost in thought and oblivious of the twelve-starred gold halo, the gold bejewelled crown, the baroque adornments and encrustations, the cloak and the roses, that have been placed about her. Underneath, she is of simple birch, and French, and about 700 years old. Her child has a cartoonish face like a smiley emoticon. His is a bit-part, not warranting any diversion of rapture away from the young Jewish woman around whom this church, this town, is built.

She was discovered (or rediscovered, as one story goes) perched in a tree surrounded by the dense brambles that choke the banks of Las Rocinas, the stream that feeds into the lagoon. A fifteenth-century hunter noticed her, dressed simply in green-moulded white linen, through the tangle of vegetation. He laboured to reach her, and finally managed to carry her away towards Almonte. He rested on his journey and fell asleep, and when he awoke, the statue had gone. He returned to Las Rocinas and found her exactly where he had first discovered her earlier in the day. Now, on the second day of Pentecost each year, a million people gather to watch her carried through the town.

El Rocío rose from the marshes at the moment of the arrival of *los gitanos* (the gypsies), whose voices were added to the resonating echoes of Al-Andalus, the ethnic melting pot that may have been conquered and converted but was never culturally extinguished. Five hundred years on, the shops purvey a unique combination of religious tat, equestrian clobber and flamencobilia. As I walk along Calle Torre la Higuera, one of the outer streets, sand piling up in my shoes, I overhear from inside an ordinary, small house someone clapping, practising their *palmas*. The radio is on, *cante jondo* ('deep song') filling the house and the street with a coarse-timbred melancholy. The voice is like a swallow's flight, flowing smooth along a line of melody, then jerking dissonantly to a new register, inflecting microtones of light and pitch, now gliding to finish a long phrase with a sudden, cut-off snap. What remains is the *compás*, and behind the shutters someone claps along with the radio: *one-two-THREE, four-five-SIX, seven-EIGHT, nine-TEN, eleven-TWELVE*: the signature of the flamenco style known as *soleá*.

I stand on the Puente del Ajolí and lean over, looking downstream along the Arroyo de la Palmosa that marks the boundary of the Espacio Natural. Beyond the bridge, a wide sand track runs for another 25 miles, passing through pines and cork oaks, and alongside *lucios* – wide, shallow lagoons. I catch a glimpse of movement in the sky, and my attention is drawn to an unruly muster of storks circumvolating a pocket of rising air. A smaller, darker shape disrupts the pattern: a male marsh harrier. His wingbeats are exaggerated, deliberately spilling air; he is signalling, demanding attention before launching his skydance. He zigzags erratically upwards, then plummets vertically towards the ground. A female appears, clearly larger, and circles the aerial arena while he repeats his dance over and over again. They cover an area of sky hundreds of yards across, hundreds of yards away; in the same view, no more than 20 yards away, a male serin rises from the top of a small stone pine, singing a fast, metallic jangle. He is also on display, also flying erratically, beating his wings out of sync to cause a wobble in his flight path. His parade ground is a circle of 20 or 30 yards' diameter, but in close perspective he and the harriers share the same visual space.

In a pause in the serin's song, I become aware of a gentle tapping sound back at the bridge. A group of small poplars and maritime junipers form a kind of wooded island in the wide sand track. I see movement to accompany the sound: a woodpecker, three-quarters obscured behind a trunk. All I can see is a dark-flecked left flank, and I cautiously step sideways to my right to gain a better angle. The bird responds by moving to its right, and my view is unimproved. I move further to the right, but I've pushed my luck, and the woodpecker flies off, disappearing into thick riverside vegetation. It is no bigger than a chaffinch, a lesser spotted woodpecker, the first I have seen in years. Despite the poor view, it is a welcome sight, not least because of what it symbolises for me, here in Doñana. It is a species that was first recorded here only ten years ago, having arrived from the Sierra Morena, 40 miles to the north. Between here and there the land is intensively farmed with vineyards, orchards and cereals, and there is only one route, a new ecological corridor, by which the woodpecker could have arrived. I played a small part in putting it there.

In 1988, I was appointed to lead the Royal Society for the Protection of Birds (RSPB)'s European Programme. My first task was to work with a small organisation, the Sociedad Española de Ornitología (SEO), to help build support for conservation in Spain. They were battling against 'Costa Doñana', a plan to double the size of the already huge resort at Matalascañas, threatening to destroy internationally important sand dunes, home to the world's most threatened cat: the Iberian lynx. The projected water consumption of the 32,000-bed resort extension and two golf courses would have sucked Doñana dry. With funds from the RSPB, SEO appointed a young scientist, Carlos Martín-Novella, who gave up his PhD research to lead the campaign in Spain, while I took the fight to Europe.

We won. After a five-year fight, the plans were definitively cancelled, and instead, in 1992, the European Commission agreed to provide 75 per cent of a €344-million investment in the local economy: the Plan de Desarrollo Sostenible de Doñana (the sustainable development plan for Doñana).

Ten years later, my relationship with Doñana had become personal. On the night of 26 April 1998, in a hotel in Slovenia, where I was chairing a conference, I could not sleep. I had tuned in to the BBC World Service at the tail end of a news story about an environmental disaster in Spain. I had missed the details, but instinct told me to fear for Doñana. 'Twelve years on, to the day,' said the BBC's Frank Smith, 'inevitably this morning's headlines all scream "*El Chernobyl Español*".' At one in the morning, midnight in Spain, I called friends. A reservoir containing waste from the Los Frailes pyrite mine at Aznalcóllar, run by the Swedish company Boliden Aspirsa, had collapsed, spilling 7.5 million tons of lead, arsenic and cadmium-laden mud, and acid water. A tsunami of poison flowed into the Guadiamar river, one of the main sources of water into Doñana, 30 miles downstream. The wave of mud and acid killed everything in the river, and spread over 12,000 acres of farmland, which will never again produce food. It flooded some of the most important wildlife habitat, killing all aquatic life and contaminating soils. Plants absorbed the heavy metals, becoming toxic to anything that fed on them.

Immediately after the conference I flew to Spain, taking with me Debbie Pain,* at the time an RSPB scientist, who was an expert on heavy metal poisoning in waterbirds, and Andy Meharg of the Institute for Terrestrial Ecology, who could advise on heavy metal transport through the soils and vegetation.

Ten years had passed between the two campaigns, and the Sociedad (now SEO/BirdLife) was bigger, stronger and Spain's leading conservation body. It would focus on

* Debbie was appointed Chief Executive of the World Land Trust in June 2016, after nine years as Conservation Director of the Wildfowl and Wetlands Trust.

supporting recovery efforts and pushing for long-term solutions while the government's clean-up was under way. I again co-ordinated international support, while Debbie and Andy worked with the Doñana Biological Station to understand and monitor the direct impact on wildlife. On the back of a campaign that challenged Spain and Europe to restore Doñana to 'better than it was before', another slug of EU investment was dedicated to a new initiative: Doñana 2005. This not only sought to address, by that date, the immediate problems caused by the spillage, but even to look at some of the worst consequences of the 1950s reclamation.

Two years ago, reporter Julian Rush and I met up in El Rocío to make a programme for the BBC radio series *Costing the Earth*. We wanted to see how Doñana had fared since we last met, in 1998, when Julian was reporting on the disaster for Channel 4 TV. As we stood on the banks of the Guadiamar, close to where the wave of toxic waste had flowed sixteen years earlier, we could see that poplars and willows had thrived in the humidity of the river and the warmth of Andalucía. The contaminated farmland has been allowed to rewild and has become a green corridor linking Doñana with the Sierra Morena to the north, hopefully reconnecting the fragmented and brink-teetering Iberian lynx population. A beautiful, pearl-grey black-winged kite hovered over the young forest. My friend Carlos Dávila, who was appointed to oversee the local response and has directed SEO's operations in Doñana since, was with us and said simply: 'Nice.'

In the 1880s, Chapman and Buck began a long and unended Anglo-Spanish relationship with Doñana. In 1882 they became joint managers of the *coto*, a 40-mile stretch of coast and its hinterland, which they ran as a nature reserve, taking what they regarded as a sustainable toll of game, until 1913

and the impending war. The hunting reverted to the Dukes of Tarifa, who, along with guests including Britain's King George V and future King Edward VIII, cared little for the long-term viability of their blood-letting. In 1952, and again in 1956 and 1957, advertising executive and amateur ornithologist Guy Mountfort (1905–2003) led a series of Anglo-Spanish expeditions bringing the expertise of leading naturalists to focus systematically on cataloguing Doñana's huge natural wealth. Spanish participants, including towering figures like Francisco Bernis and José Antonio Valverde, were among the founders of SEO in 1953. The preface to the 1968 third edition of Mountfort's book *Portrait of a Wilderness*[2] explains the British role in saving Doñana.

> Around 1960 rumours were heard that international speculators, who had already 'developed' most of the once unspoilt Mediterranean coasts, had their eyes on the southern tip of Spain ... Max Nicholson,* Peter Scott and I gathered together distinguished people, each prominent in one or other aspect of the natural sciences or the conservation of wildlife. ... At this meeting in London, in May 1961, plans were drafted for a world-wide funding organization which, in collaboration with existing bodies, would bring massive support to the conservation movement. Thus came into being a new Noah's Ark – the World Wildlife Fund.

The Doñana National Park was established in 1969, and over the years the protected area has grown to 260,000 acres, with the regional and national levels of government finally coordinating efforts in the joint Espacio Natural in 2007.

None of this has prevented a perennial sense of foreboding over the future of Doñana. Illegal water abstraction for the

* Max Nicholson (1904–2003) was President of the RSPB between 1980 and 1985.

intensive strawberry farms that start at the northern edge of El Rocío continues to exacerbate the ever more frequent droughts. At the beginning of this year,* the government finally started cracking down on water use, following yet another campaign and a warning from Brussels. If the Espacio Natural de Doñana had its origins in a British-led European initiative, then the Doñana of today is a product of the European Union. In 140 days' time I have to decide what I want my country's relationship to the EU to be. It won't be a hard decision.

The house martin is in many ways *the* hirundine of El Rocío. At the colony on the church, nests hang under all available eave space, and bracketed onto these nests are more nests. In places, the colony is three or four ranks deep.

'The swallows and house martins both arrived about three weeks ago,' says Rocío de Andrés, the young assistant in the SEO visitor centre. Rocío is from Huelva, the capital of this province, where it is common for girls to be named after the lady whose effigy was found in the dew – *rocío* – half a millennium ago.

Rocío tells me that a number of swallows are seen here through the winter. I have always had the hunch that Doñana's 'resident' swallows are really migrants, southbound and northbound birds overlapping in December and January, so that there is rarely a day on which the air above *las marismas* is not raked over by them. If that is correct, things may change anyway, and more swallows may go no farther south than here, as the climate continues to bring warmer winters. Or conversely, climate chaos may cause more summer droughts that extend into winter, worsening conditions for mid-winter swallows.

* Throughout this book, all references to 'this year' refer to 2016.

'It has been a very dry winter,' says Rocío. 'Even worse than last year, which was a bad year too.'[3]

19 February, La Serena, Extremadura. 38° 42' N. I have returned to Spain, to take the slow train to Bethlehem.

It leaves Puertollano at 11.40 a.m., heading west, and will get me to Cabeza del Buey in Extremadura in two hours – at an average speed of 35 miles per hour. After forty minutes, the first emblems of Extremadura appear: groups of Iberian magpies explode out of cork oak trees like sprays of shrapnel. They have snooker-chalk-blue wings and black caps, with pinkish undersides. They seem oriental to me, perhaps because I know they are found nowhere else in the world except the far east of Asia. Their peculiar distribution has been reasonably assumed to be unnatural – there was no record of them in Iberia before Marco Polo returned from the East. The less plausible explanation – that they were once found across Eurasia but disappeared from most of it – proves to be correct. DNA analysis shows that the Iberian birds are quite distinct, so much so as to be a different species altogether: *Cyanopica cooki*. The old name of 'azure-winged magpie' now belongs to the eastern bird *Cyanopica cyanus*.

The cork oaks' trunks are the colour of bull's liver, a sign that they have been recently stripped of their bark. Under the trees is short-cropped grass with red-earth sheep tracks winding through it. Song thrushes, crested larks and spotless starlings, feeding on the lawned grass, are put to flight by the passing train. Eventually, the driver announces Cabeza del Buey, and as I reach for my backpack on the rack above my head, I see four cranes cantering the three steps it takes them to get airborne.

I have hired a bike for the journey from Cabeza del Buey to the Ermita de Belén (or hermitage of Bethlehem), and for the three days I will spend in La Serena. Before I reach Belén, I detour north-west along a straight, paved road. After a mile and a few gentle rises and falls in the road, I stop at the next summit where, despite the occasional passing vehicle, I suddenly and happily feel alone and insignificant. The land to the left is *estepa*: stony, rolling land reminiscent of the driest downland; on my right is a remnant of a cork oak grove, but this too gives way to grass and rotation crops farther along the road. Under the trees, no more than 30 yards away, is a group of cranes. I scan the yellow-flowered meadow between the trees and count nineteen, in groups of three or four, each a family that has stayed together since their chicks were hatched last summer in Germany, perhaps, or farther north.

I take a dirt track off to the left, and after 200 yards or so leave the bike under an isolated oak tree. There is a pile of stones by the track, built where a patch of land has been cleared for ploughing. As I walk past, two hoopoes fly off from the other side of the mound. Their wings are broad and rounded like a lapwing's, white-striped on black, and seem ill-designed for any kind of flight that isn't theatrical and self-parodying, like an exaggeratedly flourished cape. When they land on another pile of rocks, they have their crests already raised, as if to goad any pursuer.

I scan the view. A flight of 200 or so calandra larks flickers in the haze; their short, woody trills, when massed, fall in the wind like a dry hailstorm of notes. Something slices powerfully through the air ahead of me, moving left. A peregrine, the most arrow-shaped of birds. I follow its flight path as it traces the skyline of the far hills, round and behind me, for more than a mile. It seems to be acting as my guide to the boundaries of its domain.

Back on the road, I continue north. Merino sheep, thousands all told, are spread through the landscape in scattered flocks, and among them, eight shapes, sheep-shaped against the dimming sky, raise their long necks at my approach – great bustards. Spain is their stronghold, with three-quarters of the world population, and La Serena is where a tenth of Spain's great bustards live, about 3,000 birds. They are a good half a mile away, and have chosen this open habitat so they can see danger from afar. Great bustards are the world's heaviest flying bird; once airborne, they fly well, but until then, could feed a lynx for a week.

20 February, La Serena. Today's dawn chorus is as sparse as the vegetation, and as thin as the chill breeze. In the sheep farm here at Belén, the first quail of the year, already returned from Africa, sings its three notes in dotted rhythm from a corner of a grass field. Crested larks utter their simple, three-note, downslurred song, neither spring-like nor lark-like in its mournful key. Corn buntings have injected some energy into the air with a jangle of notes that sound like they are forced through a sieve, pitched at first but resolving into a dry rattle. A hoopoe sends its triple-note 'hoop-hoop-hoop' across from a distant stand of eucalyptus. It blends perfectly in pitch and timbre with the murmuring sheep bells, differing only in the length of the sounds' decay.

With the sky still deeply blue and the unrisen sun casting a soft light at anything that flies at height, eight ravens appear from the south from over the Sierra de Tiros and into the air above La Serena. Like the cranes that were passing over when I arrived at dusk last night, and following the same path, they are heard long before they appear in the sky: a soft croak, deep pitched but with a high, stony note embedded in it. They have emerged from their roost in the oak *dehesas* to spread across La Serena in search of the night's casualties among the merino ewes and their new lambs. The first skein of cranes, thirty birds, appears in the northwestern sky, making the reverse journey, returning to the *dehesas* to feed. Their rough, brassy reveille signals the end of the dawn.

Cranes and ravens, two species for which the word 'icon' applies literally: like heraldic charges, there is no escutcheon more apt for them to cross than the sky over La Serena. In the strange and ungrammatical Franglais of heraldry: *Azure, cranes volant cendrée, ravens volant to sinister sable.* I try to imagine how ancient this traffic must already have been by 1231, when the military-religious Order of Alcántara took possession of these lands and their dawn skies, wielding their own green cross-emblazoned shields.

On a ridge of higher ground above Belén, a row of rocks will act as my lookout for the rest of the morning, with a 180-degree view to the north and a safe place to leave my rented bike. A pair of choughs fly almost the whole width of the panorama from my left, to an off-white stone farm building half a mile to the east. They return to cross my field of view a few minutes later, and land behind a slight rise in the ground where a group of ravens have been gathering. I cannot see what has attracted them, but the arrival of a griffon vulture confirms that it is a carcass of some kind. What surprises me is that it has attracted the ant-loving choughs. Then I realise that they are commuting between the carcass and the barn every few minutes, carrying nesting material, probably sheep's wool, from a rich enough source that it pays them to make the 2-mile round trip to gather it.

I choose a rock that is the right height off the ground to form a seat. I am facing due north, and begin to take in a half-circle of landscape ringed by the distant *sierras*, whose foothills are below the horizon and whose peaks show faintly blue against blue. There are concentric half-rings of closer mountains and hills, rock outcrops and suave undulations, and at their focal point I sit and try to gauge distances and put names to shapes. Behind me are the nearby Sierras de Tiros and de Almorchón. I pan east to west from a low, round hill topped by a red and white transmitter mast.

It is close by – just under 4 miles – and it frames the panorama to my right. Out of sight, behind and below it, is Cabeza del Buey, where the train brought me yesterday.

Moving slowly anticlockwise, my frame of view moves farther out to the Sierra del Torozo nearly 12 miles away, and where its north-west slope flattens out, the village of Zarza-Capilla catches the sun's glare. Beyond this, a few degrees farther round to the north, is the more distant Sierra de Siruela, at 23 miles. As I pan left again, I need no map to identify the castle at Puebla de Alcocer, at nearly 19 miles, with its staunch rhomboid of curtain walls and a round tower on its right-hand side.

Some of these sierras I know well, but there is an uncanniness about them seen from these rocks, like from the deck of a ship across a swell of ocean. They frame a horizon that is hidden by waves of rising and falling land. They induce egocentricity, creating an illusory circle of rock defined by the position of the observer at its focal point; they define the limits of a seen world, a personal *mappa mundi*.

I see twenty-two farms – four within a morning's walk from here, seven within a day's walk and the rest merely vague shapes in the haze. Below me the land dips for half a mile and immediately rises again to a fence line beyond a small holm oak grove, a rare heterogeneity in the landscape. This mini-valley, with a trickle of a stream at its gully bottom, is about a mile across. To get there I need only pedal a half-turn to descend a straight track the colour of an old apricot. On this side of the valley there is a strip of pasture which gives way to a broad rectangular *retamal*, around 7 acres in extent. *Retama* is a broom-like shrub, as sparse and lanky as tamarisk, which grows throughout La Serena. In a few weeks' time they will be drenched in yellow blooms. *Retamales* swathe any ground that has not been grazed or ploughed for years. They are usually straight-sided polygons, therefore, demarcating abandoned land.

I freewheel down the hill to the plough-land. I have to get off to negotiate a typical La Serena gate: a span of wire fence that can be detached from one side. I have learnt that it is easier to lift the bike over the gate, slip the top of the gate out of its wire ring and step through a triangular gap with the foot of the gate still engaged. Unfortunately, the fence is at the lowest point of the valley, and the manoeuvre must be done in several inches of accumulated rainwater.

There is a large holm oak near the gate, and I lean the bike against it to continue up the far side on foot. As soon as I emerge from its shade, two black-bellied sandgrouse appear from the left and fly past me and over the hill to my right, calling with a downslurred purr. They are semi-desert specialists, distantly related to pigeons, and hard to see on the ground, especially in the ploughed ground they favour, where their black bellies and earthen upperparts camouflage them among shade-casting stones and clods.

I walk west along the fence line for 200 yards, before turning right along a narrow headland of unploughed earth, towards a wet-flushed slade and the main stand of holm oaks. Back at the lone tree where I left the bike, a roistering chatter announces the arrival a pair of great spotted cuckoos. The first takes up a prominent position in the tree at ten o'clock, the second a more interior one at four o'clock. Their manner of swooping low and gliding upwards into position shows off the white feather-tips of their wings and fanned tail to best effect. They move off while my back is turned, and I hear them again from another oak three or four hundred yards away. They are loutish, and even the garrulous spotless starlings are spooked: thirty or so, feeding on the ground, take off and circulate a while before gathering at the top of a tree, chattering a whistled language of mutual reassurance.

A second pair of cuckoos arrive and the air carries a whiff of brewing trouble. The lapwings are nervous now. For the next half-hour the four birds thread between the trees in loud, fast, hawk-like chases alternating with a flight display of side-to-side tilting and flat-winged glides.

When they eventually quieten down, a southern grey shrike sweetens the air with a simple, calming, purling call from the top of an oak. Two hoopoes call from either side of the scene's frame. I walk north up the opposite slope, from the cover of one tree to the next, crossing 20 or 30 yards of open ground each time. The great spotted cuckoos have started up again. The starlings and other birds seem less bothered this time. It is starting to feel gently warm, and a calandra lark breaks away from its flock to make its first serious song-flight of the morning. At this signal, a woodlark rises unseen from the oaks to uncoil its helical songline. If you could draw the song of a woodlark, it would resemble a Slinky making a stately descent down a flight of stairs.

Where the ground is unploughed, it is razored by rabbits and sheep among the holm oaks and the *retama*. A yellow crucifer is in flower half an inch off the ground. In places left ungrazed, along fence lines and roadsides, the same plant may grow over a foot tall before its primrose-yellow flowers open. The same goes for the pale-purple cranes-bills and the delicate yellow-orange field marigolds. I note a strew of less ephemeral details: a squashed blue shotgun cartridge, a yellow ear tag numbered 'ES10 00017 23494' … and scattered sheep bones, perhaps the mortal remains of poor 23494.

Within the ploughed plots there are small patches left untilled, where the underlying rock is at less than a ploughshare's depth. The farmers must know these places as intimately as any coastal pilot knows each undersea hazard.

21 February, La Serena. The cranes were delayed by a morning mist, and by nine o'clock only fifty-five had passed over the *ermita*. I have decided to head west by a good track that runs alongside the railway line, rather than wait in declining hope of a spectacular fly-past.

After 2 miles, I turn right into La Serena. This is a more actively farmed area, and I ride past fenced-off fields of autumn and spring cereals, stubble, new plough and dense flocks of foddering merinos. Despite the relative intensity, there are hundreds of lapwings – winter birds here, hence their Spanish name of *avefría*. Calandra song is like a physical presence in the air, mingled into the mist. A blue tractor is ploughing; the driver returns my raised-hand greeting, while I wonder where the black-headed gulls he has attracted will have come from. Then I remember that we are only 15 miles from the Zújar and Orellana reservoirs, where the cranes sleep.

I turn back towards the railway to continue west. It is 11.00 a.m., and the sun is still struggling to penetrate the mist. There is a strange atmosphere, and the sun is a strange, muted wheat colour. Then I realise it is not mist at all, but dust – plough dust, gathered by light swirls of wind around this mountain basin and suspended at the foot of the Sierra de Tiros. As I ride past the winter cereals again, I notice a few brown, furry caterpillars crossing the track, and I slow down to avoid them. There are more and more, thousands more, and eventually I have to walk, or kill hundreds. They are all crossing from west to east; if there were any on the move when I passed this way half an hour ago, they were too few to notice. They are about one and a half inches long, and chocolate-brown with side fringes and a dorsal crest the colour of burnt sugar: *Ocnogyna baetica*, winter webworm moth.*

On the other side of the railway, the ground slopes gently, and then parabolically, up the sierra. It is fenced and wooded, a *dehesa* of holm oak, with natural oak scrub on

* Identified for me a few days later by Asociación Zerynthia following a Twitter request.

the higher ground. The sun is almost as high as it will reach today, and the light has increased because of it, albeit filtered as though through yellowish frosted glass. The birds have responded, marking a belated dawn by consensus. It is a melancholy number, though, unless something in the weird air makes me hear it that way; only the song thrush's irrepressible cheeriness seems undimmed. Everything else is solemn, sombre, sober: two woodlarks in their descending minor key, one close by, the other far away; a crested lark and a distant mistle thrush insistent but creatively half-hearted; a chaffinch perpetually down-scale; the hoopoe, woodpigeon and collared dove with their jeremiads like hired mourners.

I pass a small railway building, its roof collapsing, red bricks showing under its falling plaster, and wonder what its function was. Over the doorway, in a typographic style reminiscent of the early decades of the twentieth century, a concrete inlay is embossed with 7-inch lettering: 'Nº 182', and under that, 'Km 335.818', precise to the nearest metre. Someone has taken shots at it, leaving several old bullet holes. I continue west, where the land to the right of the track has become overgrown with *retama*. A small, concrete building half-sunk in the ground reminds me of a pillbox, but I imagine it to be some kind of storage facility, until a hundred yards farther on, a granite cross at the side of the track makes me realise that I have been cycling alongside a battlefield.

Shortly, I come to an interpretation panel on the left-hand side of the track: *Ruta Histórica 3: Ermita de Belén – Sitio de la Sorianilla*. It says:

> The pillbox forms part of the line of defence drawn by Franco's forces during the counter-offensive of August 1938. The special fortification protected both the Badajoz-Ciudad Real line and entry to the town of Cabeza del Buey. The forces deployed in the area were the two half-brigades

> of the 60th Franquista Division. They faced the troops of the 194th mixed brigade of the 37th Republican Division.

There is a photograph of the scene as it is today, with arrows to show the lines of battle. Confusingly, the panel is facing away from the view it is describing, something I slowly realise as I turn round to see the same view as depicted on the panel. It is cinematically clear: the Republicans to my left, facing the Franquistas on my right. The lines are a hundred yards apart.

> The pillbox is situated close to a cross commemorating citizens of Castuera killed by leftist militants in August 1936. The victims, who were arrested in Castuera for having links with the anti-Republican right, were taken from the train taking them towards Cabeza del Buey, and shot at the side of the railway line.

A coloured map shows the progress of the counter-offensive: 23rd of August at the Río Zújar – red; 24th – orange; 25th – yellow ochre; 26th – mustard; 27th – lemon. By the 29th, shown in primrose-yellow, the line had reached the Ermita de Belén.

I cycle back to the *ermita*, chilled by the graphic and unchanged scene, and by the knowledge that a year later, my parents' generation would witness similar horrors. A swallow flies over fields nourished by blood, feathers black in the oppressed light of a weak sun.

23 February, Alcázar de San Juan, Castilla-La Mancha. 39° 24' N. '*De las cigüeñas el cristel, de los perros el vómito y el agradecimiento; de las grullas la vigilancia.*'[4]

The train to Madrid runs through Castilla-La Mancha. Yesterday I stayed on board through Puertollano and got off at Alcázar de San Juan, in the heart of La Mancha Húmeda. The name is an oxymoron, since the first word is derived from the Arabic for 'dry land' and the second is the Spanish for 'wet'. Of hundreds of seasonal and dozens of permanent lagoons, a few remain in spite of the draw-down of the water table that has served the rapid growth of vineyards and irrigated cornfields across this vast plain. At the edge of Alcázar is the Laguna de la Veguilla, a large lagoon that nearly disappeared but has been refilled by a connection to the town's upgraded sewage treatment system.

This morning there were eighty-nine storks standing around in the field at the back of the lagoon. I imagine that they arrived late last evening, and have rested here overnight on a migration stopover. Now, at just after eleven, they are taking to the air. They split into two groups, each on its own thermal. The far group is rising faster, and the nearer birds wheel away and join them, as if thrown there by a slow centrifugal force. Four marsh harriers join in; one male uses the gain in altitude to begin a sky dance: a big dip and rise, rollercoastering with a body-twist at the lowest point in the curve. He takes six or seven dips, the last three progressively lower, until the final one drops to reed height. The female joins him for the last 60 feet of the drop, and both disappear into the reeds.

I hear cranes: thirteen have appeared in the thermal from nowhere. In their tight formation and synchronised movement, they are like a composite kite, almost rectangular, planar and beautiful. By contrast, the storks' independent motion seems amateurish. The cranes have come for only as long as they need to reach the cruising altitude they judge best, and re-form into an asymmetrical V-shaped skein of three and ten, to disappear to the north.

Miguel de Cervantes died 400 years ago this year, a day before Shakespeare. His *Don Quixote* is set among these landscapes:

> From storks, the enema, from dogs, vomiting and gratitude, from cranes, vigilance.[5]

I make a note of something to check later:

*Are white-headed ducks nocturnal? Because they're doing bugger all.**

19 March, Oropesa, Castilla-La Mancha. 39° 55' N. The N-V road into Extremadura skirts the foot of a granite hill on which the mediaeval town of Oropesa perches, presenting an image of walled self-containment and impenetrability to the wide, flat plain 500 feet below. It is a small town of fewer than 3,000, its population having peaked at 4,600 in 1950. This morning I took the 10.18 train from Madrid, and watched the cereal fields of Toledo province unroll to my right under the watch of the Sierra de Gredos and the Montes de Toledo to the north-west. After the train crossed the Guadarrama river, near Villamiel de Toledo, I was treated to the sight of ten great bustards in an irrigated field to the right of the train, close enough to be put to flight as the train passed. The slow strokes of their powerful wings made light work of the lift-off, like seeing a huge cargo plane and never quite believing it capable of flight.

In the afternoon I consult Google Earth. The eastern edge of the town is abrupt, and I find a sandy track that heads east towards a village, Alcañizo, through small farms and wood-pasture *dehesas*, between hay fields bounded by drystone walls. At the verges of the track, under the microclimatic influence of the walls, the pink of catchflies competes for attention among the fiery colours of field and corn marigolds. The catchfly, *Silene coronata*, is like a tall and delicate campion, and has its petal lobes curled inwards like

* I checked – in winter, most white-headed duck feeding takes place at night.

empty gem clasps. The small field marigolds are also in their afternoon siesta phase, orange-gold florets hinged upwards like stooks. There is no shyness about the large, yellow discs of corn marigolds, though. From the damp recesses of the stones emerge sprays of a spurge, each comprising a dozen or so stalks with saw-edged leaves and acid-green bracts and flowers – *Euphorbia serrata*.

At a crossroads in the track, by a farmstead, I pass a pond half-covered in water crowfoot, a froth of floating white buttercups that has attracted hundreds of hoverflies, which in turn are being preyed upon by several chiffchaffs, who sally from the fringing brambles. There are two small white bungalows set in three walled fields of about 2 acres in total. One field has eight old fig trees and grass, the middle one is bare ploughed soil with four fig trees and the easternmost one has tall grass, fig and almond trees. There is a movement in a bramble thicket about 20 yards away: a woodchat shrike is struggling with a small mammal, trying to impale it. The bird has its dark-brown back and its large white wing patches toward me, its head down; I can see its victim's tail and hind legs. The shrike tries another position, facing me, but deeper into the bush. With binoculars I can make out its red-brown crown and dark eye-mask among the thorny briars. It emerges at the top of the bramble, apparently having successfully left its prey to hang, wipes its beak three or four times, finds a perch – the lateral training joist of a grape vine – on which to preen, then flies off. I look into the thicket and see what looks like a shrew hanging limp over the briar, head slumped over the opposite side, back, tail and hind legs facing me, and pinned in place by a thorn.

A dry, trilling chatter causes me to look up, and the first of four waves of south-flying sand martins pass overhead, about a hundred in all. It is an ominous sign, a decisive retreat from the distant sierra back towards the valley of the Tajo. They have tasted the weather to the north, where the snow-covered Gredos peaks show clear against a sky that mixes all the colours of a woodpigeon's plumage. The sun is lost, but if anything, this coaxes more from the hoopoes, serins, spotless starlings and crowing cocks, whose soundtrack accompanies me on this walk. Among the oaks, more voices are added: a woodlark's sweet tumble of notes; a Sardinian warbler, like a short burst from a tiny, hand-cranked machine. From the top of an oak, a black-winged kite rises as I approach and glides to a more distant perch, beyond the rise in the field but just above the line of the false horizon, so that it appears to be perched on the barley; at this distance and against the gloomy sierra, it shines white.

The Sierra de Gredos, 25 miles to the north, looks closer as I arrive in Alcañizo, a trick of the still-darkening sky, whose grey is spreading south towards me, offering little hope of avoiding rain at some point. It is a village of, I would say, 300 people, and on this Saturday afternoon there is little sign of life at first. A far-from-thoroughbred German shepherd barks in a desultory fashion from behind a green gate as I walk past. Along Calle Oropesa I pass off-white houses, garages and byres. I find the Bar Albatross (spelt with a double 's' the English way), have a beer and soon leave. I notice three red-rumped swallows on an overhead cable, pass the green gate (where the German shepherd has decided he has barked enough for one day) and prepare to walk the 4 miles back towards the looming presence of Oropesa, under a sky the colour of lead. I fish a lightweight rain jacket from my backpack.

Rain. It comes about a mile from Alcañizo, where the village fields give way to *dehesa*. I head for shelter, but in the few seconds it takes to walk upwind into the trees, the fronts of my jean legs have become saturated. I stand with my back against the leeward side of an oak, listening to the rain on my hood and watching the track become a river. I wonder how long it will be before the rain stops, and when I should decide to continue walking anyway. On days like this, there comes a decisive point where the drudgery of heavy rain gives way to acquiescence – the knowledge that it is impossible to get any wetter. Then, after fifteen minutes, I notice that the flow of water along the track is tamer. The rain has become lighter. I continue heading west, and the rain stops.

I see the black-winged kite again. Against a lead and slate background it is a study in monochrome, emphasised by a slight blue-mauve tint in the sky; its wings are subtly shaded, silvery to hindward and pearly to fore, with velvet-jet shoulder patches. It has a white tail and underside, and a head of whitest grey, with a dark-shadowed eye. It hovers into the wind then begins an exquisitely controlled slow descent on a dihedral, before a final plunge to become hidden for a moment in the barley. It flies up with its prey, a vole or something similar.

20 March, Oropesa. It may be an effect of the equinox, which astronomers have calculated took place at 5.30 this morning, but I have half-convinced myself I heard a cuckoo, from a long way east. Perhaps I allowed a casual combination of pitches from within the tintinnabulation of distant sheep bells to conjure a trick of the air. I listen hard. I decide I am mistaken, and allow the bells' watery murmuring to meld once more into the sonic background.

A few hours ago, in cool air and dull light from an overcast sky, I walked through the narrow streets of Oropesa old town. I walked the same route last night, and as dusk fell I caught a glimpse of a swift as it dodged behind the church of Nuestra Señora de la Asunción. I visited the fifteenth-century castle, now a Parador hotel. On its courtyard wall is a ceramic plaque made up of four hand-painted tiles depicting a pair of the lesser kestrels who nest in its walls; I climbed the towers to watch them. The females are almost indistinguishable from common kestrels, the males separated with difficulty by the clear, unspecked plumage of their terracotta-coloured backs. In their habits they differ, these smaller birds being sociable and colonial, spending time in the castle's updrafts to play along its walls before launching out over the plains in pursuit of grasshoppers. The two species were always regarded as close relatives, until in 2002 DNA analysis showed this to be wrong. One suggestion is that as all lesser kestrels nest within the range of the common kestrel, there may be an advantage if some would-be predators confused them with the larger species and therefore avoided contact.

This morning – a Sunday – I walked around the church again, before many people had emerged into the streets. From the Plaza de la Constitución I could see four storks' nests on the church roof. By the doors to the church, a goldfinch was building her nest in one of a pair of olive trees that flank the short flight of steps up from the square. From there, I walked south out of Oropesa, along the El Puente del Arzobispo road.

Every few hundred yards, on one side of the road or the other, gates mark the entrances to farms whose names I find myself repeating out loud. Valderrevenga, La Macerrera, La Dehesilla, El Berrocal. Some names I cannot fathom, but the last two are toponyms: the little *dehesa*, from *defensa*, 'defended land'; and the land of granite boulders, *berruecos*. All the farm gates have signs saying '*Coto Intensivo*' (intensive hunting), and some have signs saying they are managed by the local agents Sprocaza.

I hear the cuckoo-effect again, two notes among many, like a half-recognised voice in a crowd. Finally, after another five-minute wait, I hear it once more, indisputably a cuckoo this time, a little closer and clearer, as if freed from a sonic cage. I note the time: astronomical spring is seven hours ten minutes old. Phenomenological spring, though, is measured by earthly signs: the pebble-strike call from deep in a bramble and glimpse of Devonian-red plumage signalling that the subalpine warblers have arrived; four red-rumped swallows hawking over wet-flushed dips in the landform; spotted rock rose, buttercup-yellow with a red-black centre.

Two great spotted cuckoos are chasing noisily through a rocky patch of scrub where a track junctions with the road. The track is tractor-rutted and puddled from rain, and runs between a ploughed field to my right and grassy fallow to my left. Its weedy edges are smoky with fumitory, and in the stony cloister of a long-derelict field wall there are five spikes of sawfly orchid. Four of them have their petals half furled like newly emerged butterflies; one is in full splendour. Its design combines insect mimicry and art deco style: the upper three petals are a lilac shade of pink, its lower lip fringed greenish yellow with a crimson-black centre. Around its mouth is a narrow ring of blue and white.

The track curves round to the north, alongside a small area of enclosed, lush pasture in which fifteen black and white cows graze and five cattle egrets patrol. There is a low, white farmhouse half-hidden in the undulations in the field. It is rough land, folded around granite boulders, with a tangle of brambles where a good thousand spotless starlings have assembled, like upside-down fruit bats, a black, wing-trembling horde. Hundreds more arrive as I lean on the wall, listening to their madding, crowding cackles and whistles. I hear a man's voice; behind me a horse is being spoken to gently and ridden with a high-stepping gait. A shallow stream crosses the track, and the horseman invites me to ford first; horse and rider follow, and turn right down a narrow green lane towards the farmhouse.

After little more than a mile, the track has circumscribed the south-west quadrant of the town, and meets a narrow road coming in from the west; I turn right and follow it back towards Oropesa. More starlings have been arriving, and the noise from the horseman's farm is now an even cacophony, like a fast, shallow river over gravel. The road undulates, and after the next rise there is a farm gate on my right leading to a rocky hillside plot that has been allowed to grow rank among stone pine and olive trees. The simple metal gate is decorated with six 35-millimetre cine film reels, a

30-millimetre gauge spanner and a huge pair of open scissors, all painted brick-red. The dip in the road dampens the starling noise and simultaneously reveals another flock-song – linnets, at least 200, which, like the starlings, have assembled for their pre-roost evensong. A few dozen serins are among them, a jangle of sopranino and soprano voices. As I watch and listen, I become aware of strange shapes among the trees. Sheet-metal flowers on steel rods, painted white and orange; a sign on the hillside saying '*Mirador de la Liebre*' (hare's lookout); a 20-foot-high pole with a toilet bowl on top and eight tools that look like hearth shovels fixed along its length; a large yellow butane gas canister in a white metal cowl on top of a pole; another pole with a cross-piece and nine more cine reels arranged in a triangle; a pole with a bedspring fixed to it at a diagonal angle and a bicycle wheel on top.

There is an olive grove on my left, which, with the waning light, is counter-shaded, darkening leaf clusters contrasting with a carpet of catchfly that gives off an illusory pink glow. I pull a leaf from a wild leek, which has a flavour like a garden leek suffused with garlic. The road rises again, marking the edge of the *finca* I have baptised Weird Sculpture Farm, and as I near the crest of the road, the blue and white pantiles of the octagonal tower of the parish church of Oropesa appear over the rise, followed with each successive step by the castle, the larger houses and finally the ordinary ones. The edge of town is marked by streetlamps and a pedestrian pavement, and by the root plates and trimmed leaves of wild leeks that someone has left on the dry stone wall at the corner of the olive grove.

21 March, Oropesa. Before I take the train north, there is time to take a slow walk out into the fields, back past Weird Sculpture Farm, beyond the point where the road joins the *camino agricola*, to the next village: Lagartera. Many of the flowers of these waysides close at midday, and I have come to take advantage of the chill morning air to get to know them better. I wander off the road and into the olive grove, where the catchflies are fully open. There are the exquisite white starbursts of *Allium neapolitanum*, known here as 'tears of the Magdalene'. There are the low spikes of powder-blue grape hyacinths, honey-scented for those who lie close enough to have the secret revealed to them, while the extrovert perfume of alexanders needs no coaxing. There are the cream-white, maroon-veined papery cross-petals of wild rocket, the white-tipped, blue-lipped spikes of narrow-leaved lupins and the airier blue of wild clary.

22 March, Alfaro, La Rioja. 42° 11' N. Alfaro is strikingly ordinary at first sight, considering its reputation. I approach from the south-east, in a car I hired in Zaragoza and which I look forward to returning three days from now. I have grown used to travelling and watching at the same time, allowing views and thoughts to drift in and out of focus. The middle reaches of the Ebro Valley – that is, the industry-lined floodplain called the Ribera del Ebro – command a driver's attention but do not reward it. On the edge of town, I pause at a petrol station called Las Cigüeñas, and spot a rusted iron sculpture at the roundabout. It depicts three flying storks, *cigüeñas*, the unofficial emblem of the town that hosts the largest single-building white stork colony in the world.

I suppose I was expecting a hilltop town with an ancient centre of narrow streets and monumental churches. I did not expect to descend to it, to drive around its perimeter wondering where among the apartment blocks and 1970s municipal buildings, factories and red-brick warehouses I would find the storks. I walk past the first block of modern buildings towards the centre of town and past a supermarket. Four storks fly overhead, their black-edged wings half-open, their legs half-extended, the classic awkward posture of a stork gliding in to land. Grateful for the clue, I round a corner in time to see the last of the four finish its glide with six or seven forward-facing flaps of its wings, braking to stall, adjusting its final approach. Their arrival is greeted by their mates with a barrage of bill-clattering from the roof and towers of the seventeenth-century Iglesia de San Francisco.

The church dominates a small, otherwise undistinguished square. At the corners of this west face are two towers, and between them the triangular gable of its single, wide nave. I count eight storks' nests – an impressive sight, but not unusual. There must be more. Opposite the church are the local offices of the La Rioja regional government, and I go there to ask for information. The receptionist refers me to a colleague, who directs me to the Plaza de España, where I will find an information centre. I find the plaza and immediately realise what the fuss is all about: a great baroque church, the Colegiata de San Miguel Arcángel. The west face occupies almost the whole of one side of the square, and is similar to San Francisco Church in basic plan, but almost twice as wide. It has the same two lateral towers, central gable and three great arched doors, but grander and more ornate, with two or three times as many bell housings. The towers are 150 feet high and in four *cuerpos* (or sections), the lower three square, the uppermost octagonal. On the opposite side of the square, shaded by columnar arches, two elderly men and a woman sit, looking across to the Colegiata and its storks as they talk.

The Colegiata stands separated from its neighbours by a perimeter passage about 8 or 10 feet wide. I walk to the right of the building, where the ground is littered with gnarled sticks of applewood, dropped or blown from above. Storks cast shadows into the passage as they fly over, and bill-clattering echoes off the walls all around me. A sign points along the passage to the *Mirador de las Cigüeñas* which I find at the end of the church compound, up two flights of steps which also serve as the entrance steps to an apartment building. On the wall of the *mirador* is the legend: '*Y nos enredamos. Ellas sobrevuelan, se quedan, viven. Y nosotros, enredados*' (And we become entangled. They fly over, they stay, they live. And we, we are enmeshed). Someone has sprayed '*y enchufados*' (and plugged in), perhaps to emphasise the contrast between the free flight of storks and the virtual lives lived by people. The *mirador* is an open, raised area 50 feet above ground level, eye level to the lowest storks' nests, which cover the roof, towers and pinnacles of the Colegiata, many of them on special raised basket-like structures fixed onto the roof. There are said to be 170 nests on the building. Some are massive, old structures, visibly layered from years of seasonal additions, like archaeological dating evidence. Many are under construction for the first time, or replacements for nests lost over the winter. There is a constant relay of stick-carriers, most with already completed nests, the apple sticks merely symbolic of their fidelity. Each bird's return is marked by a duet of bill-clattering, a sound like deep castanets, amplified by a resonating chamber in the birds' throats. This gular pouch under the chin is turned skywards as the birds draw their heads back to lay their long necks along their backs. It is a sound that in Spain is familiar and distinctive enough to have acquired its own word – *crotoreo*.

I return to the square, where I spot a green notice screwed to the front of the Colegiata indicating that there are works under way to improve management of the storks' nests (which can weigh up to a quarter of a ton) at a planned cost of €97,415.19. I wonder what the 19 cents will be spent on. I find the information centre and walk in. A woman is on the phone so while I wait I browse leaflets and brochures on La Rioja cuisine, La Rioja heritage, Alfaro walking routes, the dos and don'ts of mushroom collecting, and fishing in the Ebro. The phone call over, I ask about the storks. I want to know what the local people think of them. Do they take them for granted? Are they a source of local pride? A nuisance?

'We have a whole room devoted to the storks upstairs,' she tells me, 'but it's closed on Tuesdays; will you be here tomorrow?'

'No, I'm afraid not.'

'My colleagues know all about the storks, but they're not in today. I do too, but I'm really busy. If you email me your questions, I can get my colleagues to answer.'

She hands me a booklet about the attractions of Alfaro. It has a page devoted to the Colegiata, others to the festivals of San Ildefonso, San Roque and the Virgin of Burgo, one to local produce and, at the back, a double-page about nature. This explains that, down at the river, there is a nature reserve where the river meanders and poplar and willow have colonised. I make my way down to the Ebro, and gradually the story of the storks becomes clearer. I pass through apple orchards where winter's prunings have been left where they fell. I realise that all the stick-bringers flew in from exactly this direction. At the river, black kites and marsh harriers battle against a wind I hardly noticed in the town. Storks feed in the wet meadows, and a green sandpiper, en route to the Arctic, flies off as I appear, with a loud, rapid four-note whistle.

I look back at Alfaro and finally get it. From this direction, the Colegiata, a mile or so away, is in a direct, uninterrupted line of flight, raised above the floodplain. The Ebro is a short glide from the colony, a food source within easy reach. Potential nest material is scattered everywhere, and I have heard it said that this is deliberate: the fruit growers welcome the storks for the quantity of rodents they consume.

24 March, El Planerón, Belchite, Aragón. 41° 23' N. I have had a yen to hear the spring song of the Dupont's lark again. It is a rare song in every way: from the mouth of a rare bird, it can be heard in few places, for a few hours each day and for a few weeks only; and when it is heard, it is unlike the voice of any other bird. Its singer is seen more rarely still. It is an expert at evasion, even here, where it hides among tuffets so sparse they cover less than half the ground's surface, which otherwise is of white lichen and cracked clay.

The relief is broken by gullies and *muelas* – 'molars', the landforms left standing proud of the plain by millions of years of erosion. Nothing that lives here lives easy: on the higher ground, the gypsum plateaux, there is a plant that finds its water from within the rocks themselves. *Helianthemum squamatum* doesn't live by roots that fathom hairline cracks to find reserves of fossil dew, but by mining the water of crystallisation that builds the very fabric of the gypsum itself. No other plant is known to have this ability. Here, on these lower flat clay lands, the soil holds the winter rain for long enough to grant a living to a few arid-land specialists. Taut stems of esparto grass have emerged, 6-inch tufts showing dark, waxy green through the grey of last year's growth. The first inch-high, fleshy nodules of Mediterranean saltwort have broken the skin of the clay. In places, this skin is sun-cracked, and in the marginally more humid microclimate of the fissures, an opportunist annual, a yellow rocket or mustard of some kind, grows. Its flowers are incendiary among the dry stems, the first sparks of a brief flowering season to come. Already the first seed-pods have formed on stems whose tips still sport new flowers daily, and western dappled white butterflies flicker among them.

I hear the Dupont's call in the wind. Even on a still day it would be hard to pin down, and seems to have evolved acoustically to be an instrument of the lark's evasiveness. It sings in simple chords, diads of two main notes and their harmonic ghosts. Its main call – the one I hear now – consists of two such chords, the second higher-pitched, in parallel perfect fourths. I have no expectation of seeing the owner of this disembodied voice, but by pure chance I am looking in the right place when, for half a second, one crosses the gap between two small tussocks. I failed to see the sharp, down-curved beak that is its most characteristic feature, but I see enough of its dark, white-flecked back and its pale eyebrow stripe to know it. It is 3 yards away, and I fix my gaze on the bunch of stems where I know it is hiding. By another stroke of luck, I can approach without straying from the path onto the delicate lichen-covered ground. Two yards away, the angle of my line of sight gives me perfect vantage over his ground. I wait, but it is I who loses patience first. One yard, and it is clear the lark has gone. He has found a long, winding line of unsight the width of his own body, and made his escape.

I return to the Belchite road, and stop in a lay-by, having glimpsed an animal trotting from the left and disappearing behind a ruined farmhouse. I have my binoculars trained on the spot where it emerges: a fox, thick-tailed and the colour of pale red sand. From an information panel at the lay-by, I learn the Aragonese name for the Dupont's lark: *rocín*. Cervantes described Don Quixote's horse as a *rocín flaco*, a skinny nag, and named him Rocinante. I have no idea why the same word should be given to a lark, but in the dry steppe on a cold spring morning, were any dew – *rocío* – to form, it would seem to ring with an eldritch voice.

On a footpath that links Belchite and Belchite Pueblo Viejo, the new and old towns, fifty-nine caterpillars of the pine processionary moth march as one, nose to tail, barring the nose of the first and the tail of the last. Farther along the track, another twenty-seven are in disarray. The majority are content with their place in the line, but the first five or six cannot agree on a leader, and are moving in strange, jerky movements. Moth politics; or has a marauding magpie disrupted the consensus?

At the gates of the old town a female guide in a red jacket stands to the left, while two robust old men are on the right. One is wearing the uniform of the Nationalist soldier, *circa* 1937: tasselled *gorro de cuartel* or garrison cap, green tunic, jodhpurs and long socks, leather pouches on his belt. The other man is wearing a dark-blue pinstriped suit and dark-blue cardigan; they are arguing, disputing casualty figures.

'You need a ticket if you want to go on the six o'clock tour,' the red-coated guide tells me. I tell her that I have one, bought at the Tourist Information Centre after I booked into my hotel. There is no other way to enter the old town, which is encircled by a permanent ring of temporary metal fence panels, the kind that are erected around building sites. Around the fence, a path has been worn into the grass. I have half an hour before the tour leaves, so I follow the path to view the town from the outside.

Franco's troops had occupied Belchite until the summer of 1937, when a Republican advance sought to delay their expected northern offensive and to take the symbolic and strategic target of Zaragoza, 15 miles away. For the Republicans, the Battle of Belchite was a local success and a strategic failure. A fortnight's resistance by 7,000 Nationalist defenders allowed their generals to send reinforcements to the defence of Zaragoza. Little is recorded of the impact on the inhabitants. Perhaps there is no need – there is the testimony of the town itself: not a house is left with all four walls standing, nor either of the two churches or the monastery. The clock tower with its shell holes that might have been inflicted yesterday. The frontless interiors of homes, others with iron-railed balconies leading off phantom rooms of thin air.

I have faint memories of going with my older brother to play among the rubble piles in Cripplegate, before they were finally cleared away and replaced by the Barbican Centre. I remember the stories my mother would tell me of living through the blitz that created that rubble, stories with the terror stripped out. I think about the bombs that, two days ago, killed thirty-four people in Brussels.

I have returned to El Planerón, where in the fiery light of the evening, there is an odd truth to the false tints, like a colour-coded topographical map. No longer blanched by the sun and hazed, there is greater definition in the shapes and hues. Into the higher ground are embedded whitish strata of chalk over gypsum, while here, where the Dupont's lark sings again this evening, the soils glow red.

The wind has dropped to nothing, and as I walk I can hear the faint scrape of the unused tour ticket against the lining of my pocket. In the distance, off the edge of this protected area and SEO reserve, more virgin land is being ploughed. It is the principal threat to the lark, the fragmentation of its habitat into ever-smaller morsels. Where the available steppe is too small to support a viable population, chance events hasten their disappearance; a recent study reveals that this has happened in sixteen out of forty-two sites surveyed since 1988. There are 1,300 pairs in Spain, and none anywhere else in Europe. The fewer larks, the less they feel the need to sing. The less they sing, the fewer the chances of recruiting birds from source populations to replenish declining ones. The same study showed that in fifty Dupont's sites surveyed, rubbish dumps were found in 26 per cent, quarries in 22 per cent and wind farms in 8 per cent. In 36 per cent, they found anemometers – wind measurers, the instruments that precede the turbines.

The last Dupont's lark calls as the sun dips below the line of the La Plana hills. The cool, still air darkens. In the northern distance, a stone curlew signals the beginning of the night with a long, upslurred, yelling whistle. From the derelict farm building a little owl calls. In the near dark, the white lichens scatter like galaxies and nebulae.

27 March, Aiguamolls de l'Empordà, Catalunya. 42° 14' N.
I came here looking for nature and found two poets.

Maria Àngels Anglada described this place in the extreme north-east corner of Spain as a 'living refuge that so many wings long for from afar', a green retreat. Ten plaques, each a poem or an excerpt from her prose, form an *itinerari literari* along the trails around this Parc Natural. For two nights I have been unpicking the texts; textiles, woven from word threads by someone enchanted by her native language and the land it flowers in.

Aiguamolls is a series of wetlands once covering nearly 12,000 acres formed by the Muga and Fluvià where they flow into the Bay of Roses of the Costa Brava. It is 3 miles from the mediaeval town of Castelló d'Empúries, where I have been staying since Good Friday. There is an *itinerari literari* in the town too, featuring the works of Carles Fages de Climent, who, a generation before Maria Àngels Anglada, helped cement the modern Catalan literary heritage. On a panel in the old town, I found some lines of his entitled 'Els Aiguamolls' (The Marshes), extracted from a poem called 'Trena de set aigües' (Braid of seven waters). I copied the words into my notebook, to study later:

> *Quan l'ampla conca s'escindeix en dues / cenyint el delta obert com un parany / per escoltar el bruel que omple l'estany, / s'aiguabarregen en un sol afany / recs i Canals fent i desfent set Mugues.*

> When the broad basin is split in two / wedging the delta open like a trap / to hear the *bruel* that fills the lagoon / intervolved in one desire / dykes and canals making and unmaking seven Mugas.

I am perplexed by *bruel*. I know it as the Catalan name for the firecrest, commoner here than the equally tiny goldcrest, the joint smallest birds in Europe. But 'to hear the firecrest that fills the lagoon' makes no sense, even for a poet who was a friend of Dalí and a contemporary of Lorca. Then I discover another *bruel* associated with Aiguamolls: the Escola Bruel, a local infants' school, whose website provides the solution.

One year, in the time of Count Ponç Hug of Empúries (*c.*1264–1313), there was a poor harvest. The Count ordered that all the grain be pooled and distributed fairly among all his subjects, to protect the poorest from hunger. However, one rich farmer decided he would take all his wheat for himself. He arranged to board a ship in Roses and take his loot with him. He left in the dead of night, and took a route through the *aiguamolls*. His horse began to sink, followed by oxen, carts and cargo, and finally the man himself. Since then, the legend goes, the oxen can be heard complaining from the deep marsh every spring. This cry, known colloquially as a *bruel,* is the male bittern, uttering a deep, resonant and far-reaching booming call from the depths of the reeds.

Maria Àngels Anglada i d'Abadal was born in 1930 in Vic, 65 miles from here, and died in 1999, having moved into the Empordà, which reminded her of the ancient and modern Greek literary landscapes she studied at the University of Barcelona and whose idioms she blended into her own works. She spent time in Mytilene, whose 'wild flowers, silvery olive trees and intense blue' reminded her of the Costa Brava, 'before they destroyed it'. There is a panel at the *estany* (lagoon) del Cortalet, a shallow water body of around 50 acres, with Anglada's poem 'Els Aiguamolls 1985'. It is a rewrite of the first poem on the trail, called 'Al Grup de Defensa dels Aiguamolls de l'Empordà'. The earlier work was written in 1976 at the height of the transformation of the Costa Brava from a remote, rugged and picturesque landscape to the world's first and most notorious mass-market resort. The development had stopped at the left bank of the Muga, but plans were in place to drain the Aiguamolls and allow the spread to continue. The first poem foretells the cataclysm of the lost marshes: 'Will they invade this living refuge that so many wings long for from afar? Will the bird of the north no longer find nourishing water and green retreats?' … 'Flamingoes, our friends the mallards, farewell, farewell, Kentish plover and lapwing, colourful princess of winter.' Lines dedicated to the local campaigners who were fighting against overwhelming odds. The first drainage ditches had already been dug and bulldozers entered the marshes. The machines found their way blocked, though, by local activists, who stood firm. Children, who could not be prosecuted, poured sugar into their fuel tanks; among them was fifteen-year-old Carles Carboneras, now an RSPB colleague based in the UK.

Then in 1983, victory was declared with the formal protection of the area by unanimous vote in the Catalan Parliament. The leaders of the newly autonomous region realised the political and cultural importance of preserving this vital link in an international network of refuelling points for migratory birds. Anglada rewrote her poem, a sigh of relief: 'They have not destroyed this living refuge that so many wings long for from afar. Here the bird of the north finds nourishing water and green retreats.'[6] … 'Flamingoes, our friends the mallards, return, return, Kentish plover and lapwing, colourful princess of winter.'

In 2014 two high school students, Laia Castells and Paula Torramilans, recognising that the standardisation of Catalan bird names was at the expense of the names their parents knew, wrote *El niu dels noms: recull dels noms populars dels ocells a Catalunya* (The nest of names: a compilation of popular names of birds in Catalonia). Anglada makes a point of using popular names for birds, and I sense a gentle rebellion against standardisation in Anglada's pun on the new official name for mallard: *ànec coll-verd* (green-necked duck). Popularly it is simply *coll-verd* (green-neck); Anglada swaps *ànec* for *amic*: *amics coll-verds*, 'our friends the mallards', which she alliterates with the local name for flamingos: *àlics rosats*, preferred over the standard *flamencs*.

From El Cortalet a track leads towards the coast, through willow scrub and alongside the *rec corredor*, a drainage ditch that takes water from the pastures at the northern edge of the reserve. Anglada's 1981 short story *Flors per a Isabel* (Flowers for Isabel) is set in 1810 and relates how, in 'the first half of March … with the war [they] had neglected cleaning ditches and a large pool of shallow water lay under trees that broke into bud with their feet soaked'. These words are on a gate at the edge of a flooded field, which I leaned on to watch two angular great white egrets making geometric reflections in the still water. The trees on the far side of the field were breaking into bud, yellow-green of willow, copper-tinged lime of poplar, darker green of hawthorn. The egrets stood out like slashes in a canvas, but a steel electricity pylon blended with the unburst branches of the late-budding ash trees.

The *estany de la Closa del Puig* (Peak Fold lagoon) is a reedy pond with willow and poplar islands. If a marsh harrier were to rise from there – as one did when I walked past this morning – it would see across the Bay of Roses, where the coastline points east-south-east at Punta Falconera, and to the Serra de Rodes and Puig Pení, the peak after which the long-flooded fold is named. I had the same view from the top of an old rice silo, now a public viewing platform: it is a place to inhale the sea breeze and the scents it carries from nature, literature and history.

In the sixth century BC Greeks from Phocaea in Anatolia, the founders of Marseille, settled this landscape too. From the Greek e*mporion*, or marketplace, grew Castelló d'Empúries, while the marshes between the castle and the sea have taken a later version of the same name, hence Aiguamolls de l'Empordà.

From the old silo, you can see Punta Falconera, the headland, 6 miles distant, which might have been named by a man who robbed a peregrine's nest centuries ago. You see the surviving reach of the one-time delta of the Muga: Carles Fages de Climent's braid of seven waters; and the ditches that drained the marshes and formed the *closes* and flooded them again when war occasioned their neglect, as recounted in Anglada's story.

I descend the silo, and at the Closa del Puig watch as a purple heron drops into the lagoon. Its imperial colours seem meant only for ostentation, until they disappear among the reeds, which even under a cloudy sky still contain Mediterranean hues, reflected from below. On the path, I meet a man who is carrying a small blue cloth bag held delicately by a drawstring that has been turned once round the neck of the bag. I ask him if he is bird ringing, and he asks me to repeat the question in English, when I notice that his turquoise smock-cum-sweatshirt carries the logo of the British Trust for Ornithology. He introduces himself: Stuart Will from Dundee, and leads me along a narrow fenced-off path through willow scrub to a small stone hut no more than 4 yards square. There are a few lines from Anglada's first novel, *Les Closes*, on a panel there: 'I liked … to get to know the birds that came to live here, welcome tenants: the red-necked nightjar, the jay, the golden oriole, eater of figs, who hid away …' I learnt from *The Nest of Names* that there are many Catalan names for the golden oriole, the majority referring to it as a fig eater: *menjafigues* in Girona, *figuer*, *figueral* and *pampofigo* in west Catalonia, *pompofigo* in Lleida.[7]

At the hut, Stuart introduces me to Ian Lees from Oxford, who, like him, is spending two weeks volunteering as part of a long-term migration study carried out by the Catalan Ornithological Institute. Two other men, a little younger than Ian and Stuart and me, are demonstrating how to measure a blackcap's wing to two students. One of them turns to me and says 'hi'.

'This is our leader,' Stuart tells me, 'Oriol.' He is not the first Catalan I have met to be named after the fig-eating bird, which, according to *The Nest of Names*, is also conceited, a dreamer and a renegade.

'Hi, I'm Oriol Clarabuch, from the COI. Are you a ringer?'

'No, but do you mind if I watch for a while?'

'You're welcome.'

Stuart gently pushes his hand into the blue bag, and moments later withdraws his lightly closed fist, which has a tiny bird's head sticking out of it, at the angle between his index and middle fingers.

'Chiff,' he announces, and Ian, seated on a green fold-away chair at a fold-away table, writes PHY COL into the second column of a spiral-bound ledger. *Phylloscopus collybita*, a chiffchaff. Oriol explains to me that the project, which has been running for twenty-four years, had its funding cut three years ago, and that the statistical validity of its findings depended on a constant level of effort for ninety days between March and May every year.

'Without volunteers from abroad, the project would be over,' he says.

'Five,' says Stuart, and Ian writes 5 in the column headed *Edat* (age). I try to remember the age codes I learnt when I was a teenager, scribing for the Joyden's Wood Ringing Group. 0 – age unknown; 1 – unfledged chick; 2 – fully grown, age unknown; 3 – hatched this year; 4 – hatched before this year, age unknown; 5 – hatched last year …

'We're part of a network around the western Mediterranean. In more than twenty years, we've learnt a lot,' says Oriol. As he speaks, I hear a number being read out, prior to the ring that bears it being fitted to the chiffchaff's leg.

'What sort of things?' I ask.

'Well, we know that birds find crossing the Mediterranean much harder than crossing the Sahara.'

'How do you know that?'

'There are ringers in Morocco, on the other side of the Atlas, who catch the birds when they arrive from across the desert. They're usually in better condition than the ones we get here. More fat. They must be able to feed and rest in the desert, which they can't do over the sea.' It has long been known that migratory birds have an extraordinary metabolism that enables them to gain weight very quickly – often doubling in weight in a few days – before embarking on a long flight, if the food sources are there. The fat reserves are their fuel tanks, and are depleted just as quickly.

'Fifty … four,' says Stuart, and the measurement goes into the column headed *Ala* (wing). 'And … seven … point …

seven-point-one grams.' This is measured by sliding the bird head first into a plastic cylinder, in this case an old 35-millimetre film container, and onto a battery-powered digital scale.

'He's not a local bird,' says Ian.

'How do you know?'

'He's pretty fat, still migrating.' He holds the chiffchaff with his right hand, which is cupped over his left palm. He lifts his right hand, and for a second the bird stays resting on his left, before launching himself with an audible buzz of his wings and darting between the willows, beyond shadows.

I accompany Ian along an off-limits path to the left of the main public trail. Stuart and the students take a path through a field to the right and disappear in the opposite direction. The ground is dry but spongy, and we find ourselves in the reeds. Two nets, near-invisible front-on, are tensioned between 12-foot poles, which are fastened with guy-ropes. A ride, two feet wide by two net-lengths (about 40 feet) long, has been cut into the reeds. I hold back and let Ian approach the nets, where a single bird is lying calmly at the bottom of a bag of netting formed by the deliberate looping-up of the bottom of each lengthwise segment of net.

'Nice bird,' he says when he returns with his quarry safely in the blue bird bag. The other four or five bags, all white, that Ian stuffed into the pocket of his smock, are not needed. 'You'll like this one.' I let him keep the surprise, and he hands me the bag while he returns to furl up the nets for the day. The bag weighs no more than I would expect an empty bag to weigh.

We arrive back at the hut before the other team, and I notice that the grass around it has been thoroughly trampled during three weeks of daily ringing, except for a small, circular plot about a foot across. At its centre is an orchid, purple and erect, about 18 inches high. Ian takes the bird bag into the dark of the hut and hangs it from its drawstring on a clothes hook. As we wait for Stuart, I examine the orchid. It seems to have been saved from trampling by the instincts of the ringers, who have avoided stepping on it while at the same time appearing oblivious of its presence. Its flowers are arranged in eleven irregular whorls, about sixty florets all told. The lowest three layers have already formed green seed-pods, and the next group are a mixture of pod and withering flower. The higher up the stem, the younger the flower; those a few rows down are perfectly formed, those at the top not fully open. Each floret has the shape of a tiny pink-purple doll, big-headed, arms and legs hanging limp. The tips and extreme fringes of each pink limb are shaded with a darker tone. The three-petalled 'head', like a quiff of hair and two goblins' ears, are purple on the back, pink-flushed green on the front. I am not carrying a flower book, and will have to look it up tonight.*

Stuart returns with no birds, and sits at the table as Ian goes into the hut to fetch the final bird of the day. He withdraws his hand from the bag, and instead of making a loose fist, he is supporting the bird on the tips of his thumb and little and ring fingers, and a tiny, orange-crowned head is showing between his index and middle fingers.

'Firecrest,' he announces, and Stuart writes 'REG IGN' in column two, leaving the first column headed *Anella* (ring) blank for now. *Regulus ignicapillus*, fire-capped kinglet in Latin, *bruel* in Catalan, if not the *bruel* of Carles Fages de Climent's poem. I watch as the routine proceeds, examining a little feathered jewel I have never seen at such close range. I fail to take in the age code and measurements as they are called out and recorded, but as the bird disappears into the film container, I listen for the weight: 5.4 grams.

We have a last look before Ian releases the bird. There is a gold sheen to the green of the bird's back, like gilt enamel. On his shoulders are bright tufts like bronze epaulettes. A thin black line is drawn across each eye, joining at his black, needle-thin beak, and above that, a clear white line. His one

* Giant orchid, *Himantoglossum robertianum*.

grey feature, his cheeks, are pearly and gradated in tone, whitish fading to charcoal from front to back. The crown, a thin line of orange over a thicker yellow, bordered black, is the least outstanding of his insignia.

The clocks were turned forward last night, and my final walk back through the fields to the north of the Parc Natural is under a fresh evening sun and a shower of birdsong. A ruined farm building is marked by a line of tamarisks, from where a festive chorus of linnets swells as I approach. These are birds whose singing will not be repressed, even as they gather for their last winter roosts. En masse, their songs – individually sweet, complex and canary-like – sound like someone is slowly emptying a huge box of Christmas tree decorations down a flight of stone steps.

The evening air is warming now.

> *Always remember this light / gravid with birds almost all invisible / save for a goldfinch that swings / from the highest branch of the tallest cypress.*[8]

From far behind and high above me, a very different sound, attenuated by distance, quietly asserts itself against the linnets' concentus. There are cranes in the sky. I hear it again, closer, coming directly from the sun. Away from the glare, the sky is full of storks, seizing the last chance of some rising air to buoy them for whatever journeys they have to make in what remains of the day. I suspect that for most of them, nesting here and with no reason to wander, it is simply their nature to play the updrafts. The cranes are immediately recognisable as they drift away from the sun, even in the bleaching halo of diffracted light. Their flight formation, the synchronised planing reach that I saw in Alcázar de San Juan a month ago, is in contrast to the storks' egocentric spirals.

There are twenty-seven of them, moving rapidly upwards and northwards and into the distance, gaining enough height to see them over the Serra de l'Albera before nightfall. When they are no longer visible to the naked eye, still their cries carry and linger here. The sound has a human pitch and timbre, a feminine call that traverses moors and hangs between mountains, a throaty 'craan' with a tipped-down final cadence. It is a bleak, wuthering cry.

I continue towards Castelló, wondering if my own northward journey will bring me back into contact with them. As dusk falls and a chill returns, I hear this call again, closer this time. I turn to see six more cranes descending into the marshes to settle for the night. They may be the last cranes to leave Spain this spring. I imagine that like me, tomorrow they will be in France.

28 March, Portbou, Catalunya. 42° 25' N. The train from Figueres runs as far as Portbou, indeed a little farther, to the end of the Iberian gauge track in Cerbère, France. It twists around the coast and round (and sometimes through) the eastern foothills of the Serra de la Balmeta, which begins abruptly after the first stop: Vilajuïga. As we draw into Llançà I notice there is a higher line of mountains behind this local spur range, to the north. They will be the Serra de l'Albera, the easternmost extension of the Pyrenees. Immediately after Colera, the train enters a tunnel, and on emerging, stops at Portbou, the last town in Spain. I could stay on board, and change train and track at Cerbère for Banyuls, but I never wanted to enter France through a mountain. Still, over the last few days my plan to find a walking route over the Serra de l'Albera has been put on hold: my right ankle, and since yesterday my equally geriatric left knee, have been threatening to go on strike. But the pain has subsided today, and I leave the train at Portbou.

It has all the airs of a mountain town, a coastal town and a frontier town. The station is 100 feet above the bay and the streets lead sharply downhill. I head along Carrer del Mercat towards the shore, and look up and to my left. I see the streets on the north side of town plastered against the steep side of the Serra de l'Albera, which is reflecting the watery light with a fresh green. Unlike the railway line, which has been engineered to climb or descend modest gradients, the main road into France drops as deep as the bay, from where I can see it wending its hairpin path up this side of the *serra*. I hope to find a path over the mountain without having to walk along the road. Short, steep and wild is greatly preferable to long, winding tarmac; I suspect my ankle will agree, but not my knee. I decide on a simple strategy: keep going north, and keep going up.

From the seafront, I climb about forty steps that rise in a series of switchbacks. At the foot of the steps is a small yellow signpost with a red logo of a pair of hikers with their rucksacks and walking poles, and the words '*Coll dels Belitres 35 min.*' I guess that's the border crossing. I have to admit to a mixture of disappointment at realising that my target is well signposted and only a short distance away, and relief for the same reasons. There is also a maroon-coloured information panel that turns out to be part of a memorial to Walter Benjamin. The German-Jewish critic and philosopher was in exile in France when Paris fell to the Nazis in June 1940. He crossed these hills into supposedly neutral Spain, with a visa for America. But the Franco regime had started deporting refugees back to France and, expecting to be delivered into German hands, on 24 September Benjamin committed suicide at the Hotel de Francia in Portbou. The next day, the group he travelled with was allowed transit into Portugal and on to the USA.

The steps lead to the Carrer Sant Jordi, a steep residential street. This gives way to a narrow path alongside the final house, from whose rooftop a sweet, simple song is being uttered: three notes on the same pitch and a fourth note a minor third higher. It has a bell-like quality, and the voice ricochets off the walls. I see the bird, like a blackbird, slightly smaller and slimmer, with a longer, narrow bill. Through binoculars, its head looks blue-black: blue rock thrush. Beyond the last house, the path leads onto the open hillside, and I pause to look back over Portbou. The hills surrounding the town are steep and covered in pine and evergreen oak; as I survey the landscape, I imagine how for most of its history it must have been easier to reach Portbou by sea than over land. The neo-gothic church of Santa María stands alongside the 200-yard-long customs and immigration building at the station. The church was built by – and largely for – the railway company, and from this angle they appear as one great religio-industrial edifice. With the entry into force of the Schengen Agreement, the border formalities have been abolished, and the station buildings are largely redundant. Portbou's population, at around 1,200, is less than a third the size it was in 1930, and less than half the 1970 level.

The path over the mountain is through stony meadows with, in places, the burnt remnants of shrubby vegetation. There has clearly been a wildfire within the last two or three years, and the flowers seem to be relishing the extra space and light. There is French lavender and bugloss in the stonier recesses, and tree mallow – cerise hollyhock-like flowers with claret centres. The azure form of the (therefore misnamed) scarlet pimpernel seems to have captured all the colour that has been drained from the sea and the sky on this dull day and stored it in its petals in concentrated form. Nearby is a single clump of tall bearded iris, also deep blue. Closer to the summit, I notice the pink crumpled-silk petals and egg-yellow stamens of grey-leaved cistus, ephemeral flowers that sit at the tips of squat, grey-silvery downy shrubs.

Ahead, the path leads alongside a large white flat-roofed house, dating I guess to the 1920s or 1930s. It faces out over the slope towards the sea, and in the hillside below it, three curving, parallel lines of stone mark the contours of the slope and the ghost of an old terrace system. It marks the end of this short climb, and I have rejoined the road into France, which has wound 4 miles from the sea at Portbou to reach this point. Alongside the white house there is a smaller building, the old border post, with two small signs screwed to its wall. One, a yellow plaque, announces that this is Coll dels Belitres, and that we are at 171 metres above sea level. I feel a mild disappointment that it is such a modest altitude to have reached. The other is a signpost to the Memorial de l'Exili Coll dels Belitres, pointing to the left behind a rocky outcrop. I follow the short path between the border post and the rock, and see the memorial, built in 2009 to mark the seventieth anniversary of the Republican exodus at the end of the civil war. Rising out of a concrete plinth are four panels comprising photographs taken here between 5 and 10 February 1939 by the French-Colombian artist Manuel Moros. They depict long lines of children wrapped in blankets, exhausted refugees and defeated soldiers.

Across the road there is another yellow sign, pointing along a continuation of the path, to '*Cerbère, 4.5km, 45 min. (par sentier littoral)*', the only indication that I am in France. It bears round to the right, on the cool, shaded north slope, through a forest of pine and cork oak. The path drops sharply down to the left, where far below I see the massed parallel lines of Cerbère station and freight yard, where I will catch a train for the short journey to Banyuls-sur-Mer. But first there is a tempting detour, signposted '*Puig de Cervera 1km*'. It follows a narrow footpath along a ridge, over which a thin veil of mist is being blown from France into Spain. At the Puig I discover an old concrete trig point with a brass plate of the Instituto Geográfico Nacional. So I seem to be back in Spain. On my way back down to the main path I flush two hoopoes, which fly flop-winged ahead of me, their erratic flight paths weaving many times between countries.

France

29 March, Côte Vermeille, Pyrénées-Orientales. 42° 27' N. 'The headlands stretch out into the sea like crocodiles. In an echoing crevice of the cliff face a blue rock thrush sings.'[1]

When Olivier Messiaen came here in June 1957, he noted of the blue rock thrush song:

> [It] blends with the noise of the waves. I hear the Thekla lark as it flutters in the sky above the vines and rosemary. The yellow-legged gulls[2] cry from afar. The cliffs are terrifying. The water comes to die at their feet in the memory of the blue rock thrush.[1]

He was here as an ornithologist, and as an ornithologist he was here to listen. As a composer he had come for inspiration from the sounds of the birds and the setting in which he found them. It worked: two of the masterpieces in *Catalogue d'Oiseaux*, his collection of works for solo piano, were written here.

Messiaen's introductory notes to *Le Merle Bleu* (Blue Rock Thrush) and *Le Traquet Stapazin* (Black-eared Wheatear) describe the landscape on the coast near Banyuls-sur-Mer in detail, along with the birds he heard and whose voices he transcribed, and the impressions he gained from the colours and sounds of the sea and the cliffs. I have decided to devote this day to finding the places he describes and listening for the birds he found nearly sixty years ago.

I walk south from Banyuls along the twisting coast road until I find a path off to the left that will take me down to the sea. Between the road and the shore there is a strip of rocky heathland, and the path descends through a shrubbery of wild rosemary and feral olive, dotted with a striking yellow-flowered legume I shall have to identify later. I make some notes: *like a broom-melilot hybrid in look and habit, with leaves like huge birds-foot trefoil. Seed-pods twisted like the first coil of a helix and flattened laterally. Looks like a new pasta shape.** In places, the cracks in the rocks harbour herbaceous plants, and I notice the familiar white campion alongside large Mediterranean spurge, a robust, dark-centred species: *Euphorbia characias*. I find a giant orchid, the species I first encountered a couple of days ago in Aiguamolls, and the more self-effacing single flowers, crimson and pink, of narrow-leaved everlasting pea.

A line of reeds traces the course of a spring which forms a pool in a depression of rock. A fig tree will shade the pool in a future decade, but for now it is a new, foot-high sapling, all its energy devoted to rooting anchorage into thin soil and cracked rock. Soon I reach a headland – a trinity of them: Les Trois Moines (The Three Monks). From here I can look south along a series of other headlands – Cap l'Abeille, Cap Rederis, Cap de Peyrefite, Cap Canadell and Cap Cerbère. Messiaen's note on the title page of *Le Merle Bleu* is specific in naming the places he had come to seeking inspiration: 'Near Banyuls: Cap l'Abeille, Cap Rederis.' I am looking for a cliff face among the many minor capes and inlets that make up these headlands, using the clues he left behind. 'In an echoing rock crevice, the blue rock thrush sings … his song blends with the sound of the waves.' From Les Trois Moines I see the general area I must reach before I find the next clue.

The landscape has its signature in sound: waves against the cliff-foot; the acoustic effect of the hard rock and the soft breeze. I am struck how distance, and the angle of the cliffs, and the way the sea masks certain pitches at certain times, all affect the sounds that reach me. I hear a blue rock thrush somewhere to the south: his song varies in timbre

* Tree medick, *Medicago arborea*.

from rich and bell-like to thin and dry, not because of any change on his part, but an effect of the habitat and the atmosphere. I head south along a narrow clifftop path towards Cap l'Abeille (Cape Honeybee). Four crag martins are insect-hawking along the clifftop; it immediately occurs to me that Messiaen will have seen them, but their silence means they have no place in the two pieces that record his visit. My notebook says:

> *Gentle* clapotis *of sea on rocks. The sound has a swinging motion.*

In my haste to get something onto paper as I move on, Messiaen's own word comes to hand sooner than any I can find for myself. 'Lapping' seems inadequate when the French word has two extra consonants. When I find a rock to sit on at the southernmost of Les Trois Moines and with a view over the coves on either side, I address this sound more carefully:

> *To the left a shallow shelf of rock is washed by low waves every five seconds or so – gentle rushing, shissing sound with occasional 'pushed' or emphasised note. To right, rocks jagged with narrow strait between – complex cross-rhythm of claps and sucks.*

I look out across a calm sea that meets the overcast sky at an ill-defined horizon. There is a scent in the air, like a church the day after the thurible has swung. As I continue southwards, I discover its source when the path passes through a large patch of narrow-leaved cistus *Cistus monspeliensis*, a shrub with deep green, sticky, aromatic leaves. It is not yet ready to flower, except for a single precocious white bloom which has attracted a small, black, white-spotted beetle.* Since ancient times, the labdanum resin this plant exudes on warm

* Subsequently identified as *Oxythyrea funesta*, a species of flower chafer.

days has been collected and used as an ingredient in incense. Collection methods have varied over the centuries, from combing it from the pelts of goats that have grazed the hillsides, to dragging a special rake through the vegetation.

From Les Trois Moines, the path dips between headlands into a dry valley that leads down to a small cove of grey pebbles a few yards wide then up the other side, drifting to the right. When I reach the top of the rise I am two or three hundred yards farther inland, and the path leads alongside a series of drystone terraces that rises steeply up to the right as far as the coast road. Stumps of hard-pruned vines protrude from the bare soil, each with a unique fingerprint of whorls in its charcoal-grey bark. A male stonechat flits between vines, flicking his wings each time he lands and making a short, sharp call as stony as the soil beneath.

Above me, at the top of the next rise, an invisible bird is singing a sweet but simple song. I start up the hill, trying to avoid frightening the owner of the voice, and manage to see it at the tip of a cistus bush. In silhouette against the sky it is clearly a lark, with its short pointed crest. Almost anywhere else I would mark it down as a crested lark, but the song seems too sweet; Messiaen noted only the Thekla lark here. The two species are almost identical. The Thekla was only found to be a French bird as recently as 1931, just twenty-six years before Messiaen, without the benefit of modern field guides and the accumulated ornithological knowledge we have today, confidently included the species in *Le Merle Bleu*. Perhaps it was his extraordinary ear for birdsong that enabled him to identify it correctly.

Small squadrons of swallows have been deploying north along the coast all morning. The landform is almost fractal in the way its grand statements are composed of separate capes and bluffs, each an echo of the greater whole; each echoed in the shapes of individual rocks, and all ringing with sound-echoes whose myriad sources are indivinable. For a while I choose a path that keeps me close to the cliff edge, snaking among the many small juts of rock that combine to create each minor promontory at the fringed edge of Cap l'Abeille. The path descends and rises yet again, and the lark has moved on ahead of me. It is perched in a clump of tree heather, which is frosted with white, bell-shaped flowers, in all likelihood the ones that, by their attractiveness to bees, gave this headland its name. I turn left along a thin unbeaten path to the tip of Cap l'Abeille, remove my backpack and take out my copy of *Le Merle Bleu*.

To my left, a small cove resonates with intimate, well-defined wave sounds, and its shallow, calm water reflects a turquoise blue. To my right, the wider bay between here and Cap Rederis homogenises the sounds into a generic wash and hiss, and reflects a deep, charcoal-infused blue. In *Le Merle Bleu*, the blue rock thrush represents the sea in its different simultaneous moods. I hear one in the distance to my right. Again, the cliffs and the sea cause the song that reaches me to change constantly, varying the timbre, filtering the pitches. Only the phrasing is unaffected. Suddenly, the bird appears from between two rock stumps to my right, flies over the finger of rock I am sitting on and disappears down into the smaller cove on my left. On its short passage through my visual field, its colour changes with each wing beat according to its background and the angle it makes with the weak sun. In the score, Messiaen mentions the colour shifts of both the sea and the bird; the one represents the other. He annotates the score with ambiguous instructions: 'the resonance of rock faces', 'luminous iridescent, blue halo', or 'the horizon of the blue sea'. The harmonies are intended to complement the 'satin texture and the purple-blue, slate-blue and blue-black shades' of the blue rock thrush's plumage.

For Messiaen, there was no ambiguity. He had a condition known as associative synaesthesia, where something normally perceived via one of the senses would give rise to sensations in another. He could *see* sound – or more correctly, whenever he heard music, he *experienced* colour. Any given quality in sound consistently gave rise to the same corresponding colour sensations. Thus, for Messiaen, there was an exquisite harmony between the moods and colours of the sea, the colours and calls of the bird and the colours associated with those sounds in his head.

30 March, Le Sambuc, Bouches-du-Rhône. 43° 31' N. Riding a fast train through a slice of territory is like moving across a map where towns and roads, rivers and mountains, appear like symbols of themselves, imparting only essential information, providing a selective overview and orientation, and occasional enticement.

The route from Banyuls to Nîmes, where I will change trains, traces the Mediterranean coast and its salty hinterland for a hundred miles. The first few stops mark the gradual softening of landscape, stone-clad vine terraces giving way to fresh-harvested cork boscage at Collioure, scenes that Matisse painted in 1905, and then gently undulating plains after Elne. Beyond Perpignan, the Regional Express streaks across a broad band of Languedoc that rests barely higher than sea level. A vast lagoon lies at the feet of Salses, which, with its fifteenth-century fortress, is a thudding imposition on the skyline. On the approach to Leucate the lagoon reappears at the trackside, with dozens of flamingos like abstract gestures squirted in acrylic. After Leucate, parts of the lagoon have been sectioned off and converted to salt-works, each with its repeat pattern, a *semé* of flamingos. Small towns like Port-la-Nouvelle and large ones like Narbonne appear like castaways from their watery environment: Narbo Martius, the city's Roman name, combines the Celtic or Iberian *narbo*, a settlement by water, with a dedication to the god Mars. In Roman times it was a coastal port, but today it lies 9 miles inland, locked at the shores of its lagoon by centuries of sediment from the Aude, Orb and Hérault rivers. When the train reaches Agde the sea appears on the right-hand side, and another great lagoon, the Étang de Thau, lies to the left. Sitting on that left side of the train looking inland, there are occasional glimpses of the Caroux hills 25 miles to the north-west. But I focus in for the most part, and count about 400 flamingos between Sète and the former salt-mining town of Frontignan. After Montpellier the track curves inland towards Avignon via Nîmes, where I get off to take the next train to Arles, capital of the Camargue.

The early evening bus from Arles swings through the streets of the town's southern suburbs and outskirts, then zigzags between hamlets in the rectilinear farmland of the north-eastern Camargue. It affords a slower, closer view of these flatlands of the drained northern delta of the Rhône. In a hamlet a small, recently built cream-white house has a black cut-metal sculpture fixed to a wall. It is an elaborate cross of some kind, but the bus goes by too quickly for me to register details. Later I see the same image, wrought from simple iron bars and fixed onto the red-brick gable of a barn. The two ends of the cross-piece and the top are in the form of tridents, while the shaft of the cross forms the shank of an anchor, the whole recalling the crosses of Assyria. In the middle of the shank is a heart.

I arrive at the small hotel in the hamlet of Le Sambuc, and see the black iron cross again, a small version on the dining room wall, next to an explanatory article cut from a magazine and framed. *La croix camarguaise*, its says, was designed in 1926 by René Hermann-Paul at the instigation of the Marquis de Baroncelli. His cross is the logo for brand Camargue, said to embody the three fundamental virtues:

> The three *gardians*' tridents express Faith.
>
> The fishermen's anchor symbolises Hope.
>
> The heart represents the Charity of the Saintes Maries, the three women disciples who, according to legend, settled here after the crucifixion.

31 March, Le Sambuc. I walk across the village football pitch (which has a surface of stone chippings) to a small wooden footbridge by a sign I noticed from the bus. It is the entrance to a nature reserve, the Marais du Verdier. A narrow path leads to the reserve's perimeter track, which I follow eastwards between a thicket of tamarisk growing in the marsh to my left and a few acres of reed bed on my right. The reeds are swaying and sussurating and tracing the erratic eddies of a wind that I first noticed during the night and that has been strengthening since breakfast. The sound of the wind in the reeds blots any birdsong, not because of its overwhelming volume, but because its noise occupies all available pitches within the same broad band that birds use to communicate. Birds can filter out at least a proportion of irrelevant sound, and can hear each other despite the cacophony. Our brains do this too, but we are attuned to a different set of 'relevant' sounds, such as human speech. Today, only a bird capable of making extraordinarily high-pitched or low-pitched sounds could defeat the signal-jamming effect of the wind. And so my notebook proves: the first entry is the penduline tit, whose long, one-note call is probably the highest pitch I am capable of hearing. The second species into my notebook is the bittern, whose low bellow, at about 200 hertz, is among the deepest sounds uttered by any bird. It is produced in its distended oesophagus, a sound like a distant bull. Pliny the Elder noted that in Arles, on the northern edge of the Camargue, the bird was called *taurus*.

The bittern booms again. This bird seems to be nearby, for while it is a difficult call to locate, I can hear a detail that is usually lost over any distance – a kind of pre-boom grunt that sounds like a voiced in-breath. I stare at the spot from where the sound seems to emanate, hoping the sway of the reeds might reveal the bird for a moment. Behind me and to the left I hear the penduline tit again, and turn in time to see a small bird fly from a tall tamarisk and take a bobbing parabolic flight path, dropping into the reeds. A second bird makes the same journey, and lands at the top of a reed stem. It has a silver-grey head, a broad blackish eye-mask and bay upperparts. Its underside is the colour of reed mace seeds. They are distantly related to the more familiar members of the tit family, and differ most strikingly in the manner of their nesting, the thing that ties them to wetlands and river banks. They seek out the seed-fluff of reedmace, willow and poplar, which the male attaches to the tips of low branches to start the pear-shaped basic framework. The female joins him for several weeks of intense labour, building a hanging pouch with a short, tubular entrance. The weaving is dense, resulting in a tough predator-proof barrier that is also soft and warm. According to Finnish architect Juhani Pallasmaa, who has made a study of animal building skills, the nests were once used as slippers for little children in parts of Europe.

The *marais*, or marsh, is lozenge-shaped – about 1 mile long, east to west, and half as wide on average. It is surrounded by ditches on all sides that separate it from the village to the south, the Arles road to the east and rough pasture to the north and west. Walking round the eastern boundary and then west along the far bank, I encounter the *taureau Camargue* at close range. A *manade* of bulls is browsing the rank vegetation at the far edge of the ditch, which, I hope, serves as a wet fence between me and them. They are bred to fight, the natural instinct of all male cattle, but the Raço di Biou or *cocardier* is a fighter of men, and a very few women, in the Course Camarguaise. This form of bullfighting is bloodless, and involves men competing to collect rosettes (*cocardes*) from between the animals' horns. But the Camargue bull is more widely admired, and some are bred to be put to the sword in Spain and in those parts of France where the Spanish style of bull-baiting is preferred.

A cattle egret lands on the back of one of a group of horses grazing in the marsh among the bushes and up the bank ahead of me. They are all greys – that is, white-haired, black-skinned animals with very long white manes. None is more than about 14 hands at the withers, but they have an air of calm strength, their compact muscularity, short necks and broad heads suggesting an ancient lineage. I walk forward with some trepidation, but suddenly their calm air deserts them, and with a collective flinch they take to the water. Their hooves strike the earth with a bass drumroll, a sound that I feel in my chest and that vibrates the ground. It is a sound that would be audible a mile away, where it might be mistaken for the boom of a bittern.

'*Beaucoup d'air,*' says Monsieur Raynaud as he unchains the white mountain bike that will take me into the heart of the Camargue this afternoon. Such a wind is the least welcome of all the possible weathers the Camargue might have thrown at me. I am about to explore a vast, flat, almost treeless plain; this will not be the gentle ride I was hoping would enable me to immerse myself in the landscape, but a slog between points of interest. I set off south along the main road with the wind at my back, but I need to head west if I am to leave the intensive farmland behind and enter the true Camargue. After a mile I turn right along a lane, the Route de Fiélouse, and immediately there is a bullying response from the wind, roughing me up as I cross territory under its control. After two and a half miles of this, the dry pasture to the right of the road gives way to a lagoon, part of the Tour du Valat estate, which the Swiss conservationist-philanthropist Luc Hoffmann* bought in 1947 for the creation of a wetlands research centre.

* b. Basel, Switzerland, 23 January 1923; d. Camargue, France, 21 July 2016.

I climb the half-dozen steps to the *observatoire*, a wooden platform about 6 feet off the ground that faces across the lagoon and into the wind. To steady my binoculars, I lean my elbows on the rail, which means bending almost at right angles on splayed legs, like a giraffe. The lagoon covers about 120 acres, most of it visible from the platform. As I scan I make a rough count of the birds that pass through my field of view: thirty little egrets, eight grey herons, ten mute swans and fifty greater flamingos. Four white storks. I see a marsh harrier, and another, quartering the reeds on the far side of the lagoon, seemingly unaffected by the conditions. This is clearly a productive place, but the wind is handicapping my enjoyment of it. I get back on the bike and carry on heading west, but I have decided that instead of the 30-mile circular route I had in mind, returning to Le Sambuc from the north, I will at some point double back in the hope that the wind will have eased when I pass this place again.

After another two miles I come to the Étang de Vaccarès, the largest lagoon in the Camargue, and I stop for a quick scan, standing on the road with my legs straddling the crossbar. I count thirty-seven flamingos, but struggle to keep upright. I need to find any sparse shelter, and spot a group of trees about a mile ahead. This turns out to be the *réserve naturelle* and visitor centre of La Capelière, and the right turn between the hedges lining the entrance track brings sudden relief from the battering. As I reach the entrance door, I hear it being locked from the inside, and a hand turns a sign against the glass: '*Fermé 13h – 14h*'. I check the time on my mobile phone: one o'clock. A nature trail leads from the small nature garden outside the centre, but access is through the centre, so I walk back along the drive to the Étang de Vaccarès.

The lagoon stretches to the horizon, and its 25 square miles are thinly but evenly populated by flamingos, great crested grebes, coots and shovelers. A small flotilla of garganey take off from the water as I approach, to resettle farther into the lagoon. Half of them – the males – have a distinct white stripe over their eyes that sweeps to a point at the napes of their necks, like a diacritic. They are small – barely larger than the common teal – and will have arrived from Africa a few weeks ago, earning them their French name of *sarcelle d'été*: summer teal.

The Camargue extends across nearly 600 square miles, formed by the delta of the Rhône and by millennia of deposited silt. Eighty-five per cent is privately owned, much of it extensive ranch land where the Carmarguais horses and bulls roam freely. There is intensive rice and wheat growing, and increasingly, intensive management of water and vegetation for shooting.

Back at La Capelière I buy my ticket and follow the trail, which is just a mile long and leaves the centre through a fringe of woodland at the edge of a freshwater marsh. The linear forest of white poplars and white willows occupies a long-extinct arm of the Rhône delta, on *bourrelets* – naturally raised banks that keep the trees' roots clear of the salty water table. As a result, the forest is in the form of two parallel strips of trees, and between them the long-silted river has become a mere depression that floods in winter. From the trail I can look across the marsh to the next line of trees about 20 yards away; partially hidden at the base of a pair of willows is a vague grey shape, and as I adjust my position for a better view, I see several more. Seventeen night herons are distributed on the lowest branches, close to the water, no more than a foot apart. In their hunched-in roosting attitude they are almond-shaped: black-capped, with long yellow toes visible below their breast-feathers, and with some – the most mature birds – sporting two long, thin white crest-plumes like porcupine quills. Small frogs leap into the water, a sudden reveal as if from behind some cloak of invisibility. In places, the floodwater has encroached into the trees, and the path runs along a boardwalk. For some reason, the frogs are more approachable from here, and it occurs to me that they may be more sensitive to the vibrations in the bare ground than to the sound of my footsteps on the boards. In a leaf-lined puddle, two let me examine them closely. Although they are identical in size and shape, they are markedly different in coloration. They both have rather pointed snouts and long legs, and two ridges along their backs giving their bodies a boxy shape. Both are spotted, with about a dozen dark greenish-brown blotches on the tops of their backs between the ridges and on their legs. Both are green, but such different shades: one has a background colour of Land Rover green, the other is of sunlit grass, and unlike its duller companion, has a stripe running the length of its body: canary-yellow at the head, shading to primrose-yellow towards the back – Perez's frog.

At only 72 acres, La Capelière gives a taste of the Camargue, but no sense of its horizon-filling scale. I have enjoyed a couple of hours of sanctuary, and retrieve my bike with diminishing enthusiasm. I have come no more than 7 miles to get here, but don't relish the prospect of adding any more to the return journey, so I head back.

I stop again at the Tour du Valat lagoon. Rather than climb the few steps to the *observatoire*, I stay at ground level, where there is some respite from the wind. There is a scatter of shotgun cartridges – green, blue and black. High in the distance, a short-toed eagle is facing into the wind, counteracting the flow of air with his own forward thrust, to hang suspended in one place. He is a serpent-catcher; his body sways and his wingtips work the currents constantly to steady his position, so that he can scrutinise every bare scrape in the salt plain below. One of this morning's marsh harriers is quartering the vegetation along this near shore, a young male, more delicately proportioned than the female, but yet to fully acquire the smart blue-grey, chestnut and black livery of a four-year-old bird. He is missing four or five flight feathers from his left wing; I recognise the pattern – the bird has been shot. The spread of shotgun pellets from a gun barrel accounts for the several missing feathers, and probably means that he has taken one or two as flesh wounds. Until he grows his feathers back over the next few weeks, he is hampered. Not, perhaps, to a life-threatening degree, but he lives at the edge of the survivable world. A sudden few days of hard weather will test him more than it will test an intact bird.

And within two minutes, I catch sight of a buzzard, flying high and heading south-east. Through binoculars I see that this bird is also sporting the badge of the shotgun survivor, also in its left wing.

Under the *ancien regime*, hunting was the closely guarded prerogative of the aristocracy, and making this right universal became one of the tenets of the Revolution. This must partly explain its hold on the French psyche today: there are over 150 wildfowling clubs here in the Camargue alone, France's most important area for waterbirds. I am fascinated by this special relationship between French people – French men overwhelmingly – and the land, as expressed through hunting. From the Revolution onwards, the land has been regarded as a resource for all, and traditionally, if land is not explicitly closed to hunting, it is explicitly available. The pattern is changing, though. A quarter of a century ago, three-quarters of hunters roamed almost where they pleased, but since the turn of the millennium shooting over land leased by syndicates has prevailed. The reason is not hard to deduce: in the ordinary, undesignated land that conservationists call the wider countryside, nature has been impoverished for lover, watcher and killer alike. Had revolution come later, *égalité* might have meant a ban for all instead.

Only about 15 per cent of the Camargue is protected from shooting, and studies show that wildfowl move between protected and unprotected areas, typically roosting during the day in protected areas and moving to unprotected areas to feed. Most duck shooting takes place at twilight, either on public lands at the edge of marshes as the birds move between areas, or from hides in private shooting estates as they arrive. In the last few decades, management of these areas has intensified, with water more tightly controlled between embankments, freshwater pumped in and out of the system and vegetation stripped from banks and higher ground, with a consequent loss of biodiversity.[3] Perhaps the most telling fact is that the value of shooting leases increases with proximity to a nature reserve.[4] Here, it seems conservation benefits shooting, but it is not clear that this is a reciprocal deal.

The wing-wounded buzzard disappears into the distance, and I prepare for the final 3 or 4 miles of buffeting. I have cycled about 10 miles on level ground – hardly intrepid, but I am exhausted. A fitter man may overpower the wind, but the *mistrau* (the masterly one) is so called for a reason. I know something of how it must feel to have to beat into the storm with only one good wing.

1 April, Salin-de-Giraud. 43° 24' N. The wind is even stronger, and came charged with rain overnight; it is dry now, but there will be more – the sky looks exhausted by its burden. Under the breakfast table, my legs burn.

'*Trop d'air*,' says Monsieur Raynaud to accompany the arrival of my bread and coffee. He knows what is in my mind, and I decide that this is no day for the bike. I could walk to Tour du Valat and perhaps beyond, or I could catch the bus to somewhere and walk from there. I unfold the timetable that is still tucked into the back of my notebook, and it tells me that the bus that brought me from Arles continues as far as Salin-de-Giraud. Stretching south from there is a swathe of salt-works, the origin of the town's name. Beyond them is the mouth of the Rhône, and the marshes of La Palissade nature reserve. I'll aim for there, if I can find a taxi for the 7-mile journey from Salin-de-Giraud, and walk back to Salin in time for the last bus at seven o'clock.

The bus stops in a town square formed of rendered concrete buildings in a warm cream colour, arrayed around a broad – and today empty – marketplace. Salin-de-Giraud was created from nothing in 1856 when the Henry Merle company, later taken over by Péchiney, started producing salt for the caustic soda industry. In 1895 the town expanded with the arrival of a second company run by the Belgian chemist Ernest Solvay. A shortage of workers after the First World War led to waves of immigration: Armenians fleeing genocide, Russians escaping communism and Greeks expelled from Asia Minor. Today the square has the look and air of a Warsaw Bloc town in the Soviet era, an impression enhanced by the fact that all the shops – save the hairdresser's – are closed. One of them, with household electrical goods and assorted hardware in the window, is closed for the funeral of its owner, according to a sign in the window. He has a Greek name, and will be interred this afternoon at the Orthodox church of La Dormition de la Vierge.

It is raining again. I see an orange and white sign on a building across the square: Taxi Salinier. I walk over and try the door, but it is locked. There is a telephone number on the sign, and the call is answered quickly.

'Can I get a taxi to La Palissade?'

'From where?'

'I'm in the town centre.'

'Which town?'

'Salin-de-Giraud.'

'One moment … Is that in Arles?'

'Yes.'

'One moment … We don't have an office in Salin-de-Giraud.'

'But I'm standing right outside it. Taxi Salinier. It gives this number.'

'One moment ...'

There is a bus back to Le Sambuc in two hours' time, and in the meantime the choice seems to be between a Greek funeral in a dry church or a wet walk out to the salt-works. I head in the direction of the Rhône through a meshwork of parallel and perpendicular streets that have names like Allée des Sternes, Rue des Gravelots and Rue des Vanelles, after the terns, plovers and lapwings that I hope I will find at the salt-works.

It is 7 miles to the reserve; I would need five hours to walk there and back. I take my mobile phone from the pouch on my belt and check the time: one o'clock. I could spend an hour exploring the reserve and still get back to Salin-de-Giraud for the last bus to Le Sambuc. As I continue south, the rain is at my back, carried by a wind from the north that from time to time swirls in from the side as if for the sake of a change. If it continues, I will have two or three hours of walking into the wind this evening. In the distance I see the typical landmarks of an intensive saltscape: mounds of salt, elevated conveyors and rough brick sheds. The land is divided into straight-sided ponds by levees a foot or so high; the nearest ones are filled with water an inch or two deep, and I can see others that are dry or drying out. Sticking up here and there are the crank handles of sluice gates that control water levels to the fraction of an inch and move it around from pond to pond. There are few birds. An earthen dividing bund is the resting place for a line of yellow-legged gulls in an assortment of plumages. Hen-speckled birds in their first spring of life. Fully mature birds of over four years old wear what could be a design for a tropical-dress naval uniform of whiter-than-salt white, mantled in frigate grey, with crisp black and white wingtips like decorated cuffs. Their canary-yellow legs and lemon-yellow bills confirm their seniority. Those in the middle ranks acquire their dress in stages: heads and breasts becoming whiter, then the grey backs, until their last adolescent year, when a few brown streaks in the wings and tail, and dull legs, beaks and wing-cuffs, betray their subaltern status.

There are a few lapwings which, like the gulls, are hunkered into the wind. The same aerodynamic contours that create a wind out of still air in flight now direct the air around their stationary bodies. They may stay like that for hours. I wonder what it must be like to have a bird's mind: one that has more flight-data processing power than a supersonic fighter jet yet can tolerate standing and facing into the rain for hours. As I watch the road's surface moving backwards under my feet, I begin to face reality. The farther I walk, the greater the reward at La Palissade will need to be to make my efforts worthwhile, factoring in the return walk into the weather. And what if I slow down and miss the last bus? If I turn around now, I will get back to Salin-de-Giraud in time for the three o'clock bus.

2 April, Le Sambuc. I have an evening flight from Marseille to Manchester. The wind has dropped at last, and after twenty hours of rain, that too has ceased. At the Marais du Verdier the water is calm enough, and the sky flat and grey enough, for glossy ibises and their reflections to couple like Rorschach inkblots. There are twenty in a line, an odd kind of black: not haloed silhouettes against a back-light, nor the shining black of crow-black, but a deprivation of colour, like a line of cut-out shapes behind which lies only dark matter. Each bird has adopted a unique posture and attitude. With graphic flourishes in the form of long, curved bills and long, angular legs, they look like an alphabet. One is preening, its neck stretched down and back, its bill-tip busy under the pit of its left wing. Another has its head and bill rotated at a left-leaning angle, fossicking at the submerged base of a tussock of sedge. Two face away from me, their shapes overlapping to form one symbol. Another holds its right wing open for a few seconds. They move slowly as a group, like a line of text, gradually translocating across the mere to a new area where short, emergent vegetation shatters their reflections and destroys their symmetry, returning them to bird form.

8 April, Brière, Brittany. 47° 22' N. I cross two stone bridges, one over the Brivet where it flows through Saint-Malo-de-Guersac, the other onto the south bank of a waterway which is called Canal de Rosé on the map, Rozé on the signs, the first clue that this is a land with a multiple personality. I walk due west with the canal on my right; on my left and ahead of me is an expanse of wet meadows stretching for about 3 miles to the south. These are the water meadows of an area known as La Brière (Ar Briwer in Breton), marsh country that extends across nearly 200 square miles of southern Brittany.

There is a moment of confusion when I see twenty or so wading birds, hunched and probing systematically forward in the wet, looking nothing like egrets or anything else I might expect here. Then I remember having read somewhere that this part of France has a feral population of sacred ibises native to Africa. One flies high across my path to join the others, and I see the distinctly arched look of the ibis in flight, reminiscent of the glossies in the Camargue last week, and in Doñana and Aiguamolls earlier in these travels. Their protruding legs slope slightly down from the horizontal, as if to balance the curve of their bills. The white of their plumage has a faint cream tint, and their broad wing feathers are tipped black, banding their trailing edge. Their necks, heads, beaks and legs are also dark; in some lights the black appears navy blue. I look again at the feeding flock as the new bird joins them, and see that on the ground, they show a dark blue-black spray of feathers at the back, not easily visible in flight and from below. These are the extended feathers of the upper rump, lax-plumed like those of an ostrich.

The afternoon weather is unsettled, and to the south-east the sky has turned the colour of fresh lead plate after a few days of exposure: dulled-silver-tinted petrol-blue, blotched with dark pearl. Farther along the canal the scrub hedge thins out to nothing and the landscape to the north opens out, while ahead the canal enters the flooded marsh and the track peters out into the wet. At the edge of the floodwater is a group of willows that have been pollarded during the winter, and are yet to sprout new growth; one trunk has an exposed cross-section the shape of a perfect heart, waiting for two sets of initials. Out of the corner of my eye I spot a movement – a small bird appears on top of a willow but disappears into the base of a patch of reeds as soon as it sees me. Robin-sized, with a cocked tail, but in silhouette: still, enough to recall the shape that I saw one morning in Doñana. More than two months have passed since then. The bluethroats of southern Spain will have moved north, and perhaps this was a brief glimpse of a fellow wayfarer travelling the same route as me. I can go no further along the track, and when I turn to retrace my steps to the village I see a couple holding hands and heading towards me. They are about sixty years old, both wearing red waterproof jackets. I toy with the idea of lending them my pen and pointing out the heart-tree, but I am distracted by a skull, a coypu whose 2-inch-long orange teeth are half buried in the waterlogged soil at the base of a neighbouring willow. It has remnants of sinew and skin attached, and I ease it up onto the surface of the ground with my shoe. The teeth are bevel-tipped. In life they grew constantly, and the difference in hardness between the orange outer coating and the whitish inner layers ensured they remained whetted, grinding differentially to form a self-sharpening chisel-tip.

I drive north in the hope of finding better access into the marsh, and after a mile or two, I see a sign pointing left, to

Île de Fédrun. Islands in marshland invariably have a rich history and distinct character, so I turn along a road which is raised above the water in the surrounding reed bed but which must once have been a low, seasonal causeway into the mere. The village of Fédrun is less than a mile along it, but as soon as I arrive and park the car, I sense a separation and a remoteness, perhaps more personal than geographic.

Swallows are here by the handful, not by the hundred as they were in the Mediterranean a week or two ago. Whatever governs their abundance in these early days of spring, it is not journey time. If it needed to be, a swallow could be in the Camargue one day and hawking over the Brière one or two days later. There is a taste of the Atlantic in the air, of an indecisive spring, where the scents are of distant rain and blackthorn, and no more of cistus and dust and rosemary. Lapwings are here to sky-dance their territories, not to wait out the northern winter.

Fédrun itself is gently (and genteelly) busy with knots of middle-aged walkers, teenagers hanging around on bikes, gardens being tended and a *crêperie* at the very edge of the marsh, where the causeway rises a few feet onto the *île*. I pause there, watching marsh harriers and a black kite over the reeds as I wait for my crêpe. I am facing back towards the eastern shoreline of the mere, the mainland-in-the-round to this island, where the village of Saint-Joachim is mostly hidden among the poplars, save for a few white gables that show through the green-hazed but unleaved trees. Mistletoe clumps are scattered through the crowns of the trees, like clusters of green nebulae.

I ask Jacky, the owner of the *crêperie*, how long it would take me to walk round the island, and she tells me forty minutes. She has installed posters of the local birdlife on the terrace, along with a panel with over a hundred symbols on it: ducks' feet, each pair uniquely marked with nicks and slits and identified with the name of their owner. She sees me studying the flag that flies at the corner of her property, at the entrance to the village off the causeway.

'*C'est le drapeau du Pays de la Grande Brière,*' she explains, and goes on to describe the design: divided in four behind the black cross of Brittany; the first and fourth cantons green with the gold figure of a marsh duck; the second and third white with eight flecks of ermine. It was created in the 1920s by René-Yves Creston, *briéron* artist and ethnologist.

I pass a terrace of four stone-built houses, the second of which has an A-board outside advertising itself as a museum – free entry. I walk through the open, green peeled-paint door expecting to be greeted by a curator or receptionist, but there is no one there, and no desk or chair suggesting there is ever anyone there. It is an unlit single room, the downstairs part of the house opening directly onto the street. The walls are lined with tools and artefacts of *briéron* life. Various sizes and shapes of peat-digging spades lean against the left-hand wall, along with a toothed eel-gig, like a spear with five flat serrated blades to hold an eel fast. A long, narrow, shallow black wooden box about 6 feet long leans on the front wall. It has the words '*vivier à anguilles*' painted on it in white – a live-box for eels. There is a pair of black leather boots and an oak bedstead with a linen pillow and duvet. On the back wall is a cabinet of stuffed birds, next to which a sign explains that they were taken in the marshes before they were protected by the law of 1976, and that they are absolutely not for sale. Nothing is explained, almost nothing is labelled and nothing is secured from theft.

I take an anticlockwise route along the road that encircles the island. There are houses on both sides of the road, and from a postcard I bought at the *crêperie* I know that the convex centre of the island is covered in unhedged pasture and hay fields. The aerial view makes it easier to understand this landscape. The white houses that encircle the island look like a necklace of white teeth thrown casually into the marsh to land in a rough rectangle. They in turn are bordered by an outer palisade of trees, which are themselves bounded by a kind of ring-canal – the *curée*. At first sight, the island looks to be set in a huge reed bed; it is, but the postcard clearly shows ditches cut through the reeds in parallel lines. It is the signature of former peat diggings, rectangular reeded-over basins with deeper, navigable channels alongside, known locally as *piards*. I pass between the double row of houses. Each of the outer, marsh-side dwellings sits in its own long, rectangular plot that curves almost imperceptibly down into the marsh, like a meniscus. Many of the landward houses have long vegetable plots on higher ground. Their range of style, size and state of repair seems infinite. Most have thatched roofs – some recently re-roofed, most mossed-over to varying degrees of green. The older houses are of unrendered stone and single-storey, while others are white, two-storey and slate-tiled. One small cottage has a thick thatch of reed covered in decades' worth of moss, exterior paint peeled half away and a satellite dish on the front wall. A barn-like building has a concrete front gable, *circa* 1940, fitted with a modern garage door. Its reed roof is invisible under a forest of polypody ferns and wall pennywort. Its side walls are crumbling, exposing wattles of spokeshaved alder. Many of the houses have elaborate triangles of crocheted lace in the windows – *rideaux brise-bise* – each as apparently unique and intricate as snowflakes; and all have blue-painted window shutters.

A sign, '*Vente d'Anguilles*', lies on the ground in a garden, waiting for the eel season. The next plot is vacant, and I walk through it to the wet edge of the *île*. The edges of the plot, to right and left, comprise straight ditches cut into the peat, 6 feet wide, and filled with water. A punt has been drawn up onto the edge of the right-hand ditch, under the island's ring of poplars – some years ago it seems, as it is now full of leaf mould and lush nettles. A green woodpecker appears from the far side of a poplar trunk, and I hear the intermittent flutter of its waveform flight between the trees. Between showers, the sun has been getting brighter towards evening, and washes an orangey cast through all the greens – the woodpecker, the budded poplars, the grass and the wall of reeds at the edge of the *curée* that bounds the view from this low vantage. Behind me, to landward, I hear distant strimming. There is a second green woodpecker somewhere, and the two birds call against each other, a series of strident yelps falling microtonally in pitch. And another sound, mysterious and primordial, as if upwelled from deep in the silt, perhaps somehow self-emanating; or the delayed echo of an aurochs call; in any case, a begetter of tales from the deep marsh: a bittern.

9 April, Saint-Lyphard, Brittany. 47° 24' N. On the high ground to the west of La Brière is a landscape of scattered villages, small farms and woods. The morning sun has prolonged the dawn chorus, and I follow a track through an oak wood to the sound of robins, song thrushes, chiffchaffs, blackbirds, nuthatches and wrens. I listen for willow warblers, my personal herald of spring. This is the date I have always associated with them in the North of England, despite my own records showing that they have been arriving ever earlier with the changing climate, in some years before the end of March. So I have been expecting to hear them here in France, some well settled on territory, along with northbound birds passing through; but I hear none today.

The land undulates gently, in ripples rather than hills, forming linear depressions, in one of which is a reedy river: the Mès. Cetti's warblers hurl their notes like rapid-fire arrows, in short bursts as if reloading their quivers between volleys. They are almost always hidden deep in the reeds, but as I approach the water's edge, one sings loudly from a willow bush at head height, about 15 yards away. I have reached the Pont de Gras, *monument gallo-romain*, a 2,000-year-old stone causeway comprising about seventy flat-topped stones, worn smooth and averaging about 2 feet square, laid across the shallow river bed. I walk across and back, stopping halfway, where the bridge crosses an open ride through the reeds. Spraint, which from the absence of fish scales and its carnivore smell I presume is of mink and not otter, has been deposited on one of the stones, a mark of territory. At one end of the causeway an oak tree has rooted itself among the slabs and built a skirt of buttresses across them. Its roots are half in water, indicating that the river level is temporarily high, but that normally – or at least seasonally – the tree is in no danger of drowning. Its trunk is green with vegetation: about two and a half feet above the ground is a near-perfect separation of moss below and lichen above, like a tideline.

A cuckoo calls. In my notebook I write: *slightly sharp minor third,*[5] having decided this spring to test the view (which I have long held but never systematically checked) that the minor third version of the cuckoo call is most common in early spring, and that later the major third predominates. It is the difference between the cuckoos of 'Sumer is Icumen In' and Beethoven's *Pastoral Symphony*.

10 April, Forêt d'Écouves, Normandy. 48° 33'N. I stand in the shadow of a holly bush, pretending to myself I am inconspicuous. I hear a sharp call to my right; a firecrest has come to investigate me from 4 feet away. I can see into his tiny mouth as he calls, and admire the design of his face and crown, with his badger mask and flaming-orange stripe like a vivid centre parting. I catch sight of a treecreeper about 25 feet up a beech tree, about 20 yards away, and see that it has a pure-white underside, making it the Eurasian kind, normally the only treecreeper that is seen in Britain, but one of two species found on the Continent. I also hear the simple, lilted song of the other one: a short-toed treecreeper. The two birds are about five trees apart, causing me to ponder what subtlety in their ecological niche keeps them from competing with one another. In the distance I hear a green woodpecker's sarcastic laugh, and the insistent piping of a nuthatch. Life here, at this juncture of winter and spring, is lived on the lichen-rich trunks of the trees more than in the leafless crowns.

This forest is 1,000 feet above sea level; it has the look of winter and the sound of spring. The birds are responding to the lengthening days, a phenomenon exaggerated by the horizon's position, far below. It is mainly the local birds giving voice; apart from the ever-early chiffchaffs and one or two blackcaps, the summer birds are nowhere to be heard – no cuckoo, nor willow warbler, so far at least. At this elevation the beech trees, like the oaks and hornbeams that are scattered among them, are their winter grey. Their tight-fist buds may number millions, but together they barely even hint at colour. Dark Scots pines and hollies show as shadowy cowled figures in the ashy mist of bare twigs and branches. Look closer, though, and there is a veneer of life throughout the forest. It is the damp, bright high summer of the epiphytes and lithophytes. Smooth rocks are planets in themselves: jutting proud of the leaf litter, they are forested by *Polytrichum commune*, haircap moss, like primitive conifers towering inches above the surface. Roaming among the dark, wet fronds are countless and nameless beetles and spiders; tiny to my eyes, they are the voracious beasts and timid beasties of the micro-jungle. The trunks of the trees, for the first 5 or 6 feet, are cloaked in another moss, *Thuidium tamariscinum*. One of its English names, fern-moss, is apt. At another scale, it could be a steep bank of bracken on a sun-facing escarpment. These few weeks of expanding days, softening chills and an absence of leaves to cloud the sky is their time.

The beeches and oaks are made tall and straight and are almost all unbranched below the crown. They bear little resemblance to the stout, lateral branch-bearing trees of a wild forest. They rock to the command of the breeze, and shout and cough and drum, when a Sherwood oak might creak a soft complaint once in a while. Finished timber, trunks cut into identical lengths, is stacked in flat-topped pyramids by the side of the track. Each bears marks along its length like lizard tracks on damp sand, where a forest harvester has gripped the cut tree, de-limbed it and fed it into the maw of a hydraulic power-saw: from living tree to stacked timber in a few seconds. As I walk past, each pile gives off its faint and distinctive scent which, in this chill, needs a wisp of breeze to trouble the air if it is to be noticed at all – except for the newest spruce, whose resinous perfume tangles promiscuously around the forest's vapours. Old beech, I notice, has already acquired the smell of furniture, as if the tree were made from the table.

I return to the car and head towards Goult, a place I had never heard of prior to searching on the internet for nearby Natura 2000 sites – the EU-wide designation for areas protected under the Birds and Habitats Directives. Emerging from the forest, along a quiet lane, there is colour in the hedge-banks – celandine and wild daffodils. I spot a pair of roe deer grazing in a field, but as I slow down to stop they bound into the wood. I watch them moving about the trees, their presence only occasionally betrayed by a sight of their off-white rumps in the thicket. Several minutes pass before I realise that they are no longer there. Like swifts in high summer, their leaving is known only in retrospect.

I arrive at the Tourbière des Petits Riaux. It is a miniscule remnant of bog-land in the midst of a birch-wood. There is just enough space for a tree pipit to parachute its song into. It is loud, proud and exuberant: launched from a perch, the song begins on the up-flight of its near-vertical rise into position over the arena. Each jerk of effort is accompanied by a short, sharp call in an accelerating series. When he has reached the required altitude, 50 or 60 feet above the ground, he descends on fixed, spread wings and cocked tail, adjusting the angles of his wings and body to slow the pace of the fall, allowing the air to move him around in a loose spiral. His notes coalesce into a trill, until, 20 feet from the ground, he takes control of the final glide to his perch and finishes the song with three stretched and slightly tremulous notes. Perched, he barely registers any visual impact; he has the small, brown, streaky default design of the standard bog dweller. Evolution has chosen crypsis over colour for him, and compensated with a voice to infuse the air. His singing is incessant. This is the only acre of restored bog in the district, the birch having invaded decades ago when centuries of grazing and mowing came to an end. He needs to let others know he has arrived: to attract a passing female, and to let any rival males know they are too late to register their claim.

His positivity seems in contrast to the dark and claustrophobic air hanging over the bog, as if he felt duty-bound to inject notes of hope and optimism. Downy birch trees surround the space as if laying siege, and will encroach as soon as one season passes without intensive scrub-clearance. The truth is, this is not so much a nature reserve as a small open-air museum. There is a boardwalk snaking through it, and every 2 or 3 yards an off-white information label has sprouted on the end of a 3-foot-high mild steel stalk. There are twenty-five in all, each identifying one of the special plants of the *tourbière*. Most of the plants are several weeks from flowering, but at this early stage of spring I am happy to know that they are there, somewhere, summoning reserves from within the peat for the time of the bog-lands' brief and understated glory. There are two species of cotton-grass, the fluff-headed sedge of the wet, nutrient-deprived peat; and sphagnum, of course, the main ingredient of peat itself. This family of mosses plays a critical role. Each plant grows continuously, leaving its lower parts to die and compress over thousands of years while the upper few inches strives continually for light. There is round-leaved sundew, whose carnivorous habits enable it to obtain nutrients from the local insect population, the waterlogged peat's nitrogen being locked away from any plant that has not evolved some way to fix, steal or replace it. There are the bringers of fleeting colour-joy: bog asphodel, heath spotted-orchid; and of majesty – the royal fern, its fronds as tall as a man.

At the village of Goult I stop at a small chapel with a three-order Romanesque portal, reminiscent of the rounded

Norman arches in old churches in England. The Priory Chapel of Goult is a few yards square, and stands on its own grassy mound at the edge of the village. At the top of the six portal columns are carved capitals, each featuring a twelfth-century depiction of a scene from the natural world, succinctly described on a nearby information board in French and English:

> A bird struggling with a four-legged animal.
> A bird pecking at a twisted twig and two lions fighting.
> Interwoven birds and a small animal in the foliage.
> An archer walking towards a deer with a knight sounding a hunting horn.
> A hunter brandishing a spear to kill a wolf at the throat of a goat.
> Some birds and a fox.

The carvings seem more than just a mediaeval stone bestiary, and in part also a document of country life. This area is known as La Lande-de-Goult. *Goult* is an Old Frankish word for wolf; *lande* refers to moorland or heath, the last fragment of which I have just left behind. When the last wolf was killed in the surrounding Forêt d' Écouves in 1882, the population of this village was 450, from a peak of 600 thirty years before; it had declined to under a hundred by the late 1990s, when the abandoned *lande* had been almost completely taken over by forest, and the project to restore fragments of it began.

I find another Natura 2000 site, a small wetland, the Marais du Grand Hazé, on the edge of a small town, Briouze. On the way to the *obervatoire ornithologique* there is a large sign with details of hunting seasons, management responsibilities and fishing arrangements. Inside the hide are various phone numbers and scribbled messages that fall outside the scope of my French vocab, but from its window slits I am able to see the comings and goings of a heronry of twelve nests. On the water there is a coot and two mallards. Two buzzards circle overhead, one with broken primary feathers, like the shot buzzard I saw in the Camargue. I spot a small flotilla of waterfowl: two teal, a great crested grebe and five decoys, which are spinning slowly in the breeze.

It is getting late, and I have a journey to make, so I decide that this place is not for me. I am yet to be persuaded that the accommodation between hunting and conservation in France works to balance the desires of one part of the population with the common heritage of all. I look again at the information board before getting into the car. Most of the reserve is divided into equal halves, with the right to shoot split between two clubs, along with a small number of private shoots. Management duties are shared between the shooting interests and the local conservation group, based on agreed objectives. The notice does not state what these are, but protection of the 'cultural heritage' (code for 'hunting') is usually one. Management, it says, includes releasing captive mallards before the hunting season; local by-laws determine how many are to be shot and how many are to contribute to the breeding population.

11 April, Le Champ-de-la-Pierre, Normandy. **48° 36'N.** I enter a dark, dripping, still, ringing forest by a lake. It rained heavily in the night, and under the trees it is still raining, hours after the rain stopped falling; and the morning birdsong has percussion to accompany it, the slap of water on last autumn's leaves.

The forest has been managing itself for several years, as I can tell from the fallen branches that lie everywhere. A thin branch of beech has been colonised by overlapping scales of a fungus, a so-called resupinate, the whitest thing in the forest, the colour of ivory and with a false glow conjured by the gloom and by my eyes' over-compensation for it. I follow a remnant track, a shallow shadow in the leaf litter that has seen no vehicular traffic since an old holly fell across it years ago. There is one small pile of timber, mossed-over and rot-infused. Bramble, honeysuckle and wood spurge form a ground layer, threadbare in places but everywhere more luxuriant than in the forests I walked in yesterday.

I arrive apparently unnoticed by a troupe of woodpeckers too preoccupied with their trunk wars. From 50 feet below, it is like watching an ancient and unintelligible board game in three dimensions. It takes me several minutes just to ascertain the numbers: that there are six great spotted woodpeckers and a pair of middle spotted. After half an hour, I have deciphered some of the play: two of the great-spot pairs are laying claim to trees around me; their territories overlap, but not entirely. The third pair live mainly in the adjacent territory to the west, with little overlap. The middle-spots live mainly to the south, but spend time within my field of view. Great-spots chase each other incessantly with a loud, grating call; male chases male, female chases female. The western pair enter the field of battle to chase back any rival that strays their way; then they stay and join the fray for as long as they deem necessary to preserve the integrity of their own domain. The middle-spots observe, like correspondents, following wherever the action takes them. Otherwise, they focus on each other, signalling their secrets with shaggy, raised crests.

The great spotted woodpeckers don't have the red, expressive crowns of the middle spotted; instead, the male has just a small red patch at the nape. Recent research in Poland[6] suggests that middle spotted woodpeckers signal more than just affection with their red crowns. The colour is acquired afresh with each annual moult, and its brightness is determined by the quantity of natural pigments in the birds' diet at that time. These carotenoids are produced by plants, whose colour persists through the food chain. It is what gives the flamingo and the salmon their pinkness, and puffins their ornamented bills. They are antioxidants and immunostimulants, but depositing a proportion of them in plumage is at the expense of these health-giving properties; and is costly in other ways – absorption, metabolic conversion, transportation and incorporation into feathers. Pairs with the brightest, shiniest red caps produce more young, from smaller, better territories. Their red caps are described as an 'honest signal' – a reliable sign of quality (as opposed to the 'dishonest' signals given by harmless insects that mimic harmful ones). Middle spotted woodpeckers with the shiniest caps were found to be in better body condition, and were able to spend precious time and energy grooming, to keep up appearances. The best-conditioned birds of both sexes use these signals to seek each other out, as likely perfect parents.

I was on my way to the marshes in the floodplain of the Seine when the dark wood invited me to stop and feel the rain I missed in the night. I am delayed again, by a sudden emergence of the sun, which sparks a blaze in a bank of wild daffodils. Near Ménil-Jean I stand on the bridge over the river Orne and look down. The water flows steady, without breaking white. Below the surface, long tresses of a green weed wave in the liquid breeze, like the hair of a Pre-Raphaelite Ophelia. In the air between the river and me, small tricks of the light, like grains of mist, are rising by fickle routes through the air. In minutes, their number swells from a handful to hundreds, and I begin to notice them glistening in spiders' webs strung between the galvanised steel railings of the bridge. Only by halting their motion are they given substance: caddis flies; their wings are diaphanous, translucent and faintly marked with grey-green spots.

The margins of the river are narrowly wooded, and at this lower altitude the banks are a Milky Way of willow catkins and blackthorn flowers. Among the trees I find yellow archangel and violets, and in a hay meadow beside the road, pink lady's smock and yellow marsh marigold. It was the wild daffodils that made me stop here, and I crouch by the hedge-bank so that I can look up through the sward and see them back-lit. There are thousands reaching out from the spring grass, their stems angled by the gentle weight of their rain-washed flowers. I am left wondering what is gained by breeders' continual striving to reinvent *Narcissus*, when all that can be achieved is to lose its delicacy of stature, saturate its subtle yellows and caricature its manifold forms.

Marais-Vernier, Normandy. 49° 25' N. I have found a place to stay on the banks of the Seine, where the river forms the wide meanders or *boucles* of the Parc Naturel Régional des Boucles de la Seine. It is a wet evening, and I drive through a wooded part of the *parc* to the Marais-Vernier, an 11,000-acre depression tucked into a loop of the river made up of wet meadows, reed beds, meres, peat bogs and *curtils*, from which we get the English word 'curtilage'. These are ancient strips of once-cultivated land that have become a habitat in their own right.

The easiest place to gain access is a tall observation tower at la Grand Mare, a large area of open water, part of a hunting reserve managed by the Fédération Départementale des Chasseurs de l'Eure. I climb the wooden steps towards the covered top platform, where a couple in their early forties are peering through their telescopes. Before I reach the top, the man looks down at me and says:

'*Balbuzard pêcheur*!' and then, 'osprey!' We watch the bird as it quarters the lake, high above the water, perhaps as high as this tower, and perhaps two or three hundred yards away. Now and then one or other of us has our attention diverted: a fat coypu, a swallow, a dog barking at the pinioned ducks on the small pond below; always returning to the osprey.

The conversation takes the form of a running commentary in two languages: *I've lost it – It's over the pylon, flying right … bloody dog! – Swallow, left – Where? – Left, here, very close! – Where is it now, oh I see it – Got it? – You'd think they'd keep the bloody thing under control – Is that a coypu? – Quick! It's going to dive! – Where? Oh yes, got it, oh! It's dived … it has a fish!* The man's English is quite good, and my French is, too. The woman speaks relatively little, all in French.

Only when the osprey has taken its fish to a perch beyond our view do we introduce ourselves. Alexis and Christine Nouailhat are on holiday, mainly birdwatching, and have driven here today from their home near Grenoble. They will head for the coast this evening and find a place to stay. I mainly speak in English, and Alexis speaks mainly in French. I ask why Alexis told me about the osprey in both languages without knowing I wasn't French. I was more likely to be foreign, he says, coming to an observation tower with binoculars. We discover that we are both professionally involved with nature, I as a conservationist, he as a wildlife artist.

'My parents spent a lot of time outdoors, looking at nature,' says Alexis. 'So I've been birdwatching since I was ten.'

'I'm fascinated by the fact that so many nature reserves are managed by hunters, like this one. How is the relationship between hunters and conservationists these days?'

'It has been a big problem in the past, but it works better now. There is more collaboration.'

Christine makes a facial gesture, an international signal of scepticism, and with a slight tilt of his head and flick of an eyebrow, Alexis seems to signal that Christine may have a point. I ask about French people's interest in birds and nature, and Christine, who has said little so far, says, in excellent English:

'When people think of nature they are mainly thinking about hunting. Even in strict nature reserves they focus on management for shootable species.'

It is time for me to move on. I ask Alexis if he has brought a sketchbook, and if I could take a look at his work.

'In the car – I'll come down with you.'

He opens the rear hatch door of his car, which forms a shelter from the rain. The back seats are folded forward, and along with a small quantity of luggage I see an easel, boxes of paints, piles of prints and a shoebox full of postcards. Alexis pulls a postcard from the box and hands it to me, a cartoon of dancing cranes in the snow. I recognise the style – humorous, and just stylised enough to be fun while remaining essentially accurate.

'I know your work,' I tell him. 'I've seen it on the cover of *L'Hermine Vagabonde*.' I came across the magazine ('The Roving Stoat' in English) in Brittany, where it is produced by the organisation Bretagne Vivante. It's a children's nature magazine in which the stoat is the reader's guide to the natural world, exploring its subjects in depth and with scientific rigour using Alexis's cartoons and humour.

'Yes, I've done all their covers and illustrations for many years,' he tells me as he hands me another card, a little ringed plover sheltering her chicks from the rain.

'For you,' he says, and he hands me another, then twenty more, one by one. Then he reaches into the car and flicks open a fat sketchbook that is lying on a suitcase.

'Watercolour. I don't want to bring it any nearer the door in case the rain touches them.'

It is a village landscape, and comparing it with the postcards I have in my hand, I see they have something in common, despite being in a completely different style. Whether capturing the anthropomorphic glee of a kingfisher as it grabs a surprised-looking minnow, or the rustic dishevelment of a farmhouse in the mountains, he is a close observer of light, of all its qualities and effects. Of my collection of twenty cartoons, only one is set in the perfect light and shade of a bog-standard sunny afternoon. The red-breasted geese are under an overcast sky; the waxwings are in the murk of falling snow; one card shows Arctic sea-ducks at midnight: the summer sun is low, hidden behind the mountains, but wields a strange influence over the sky, the snow and the sea. It is a light I have never seen, but hope I will next month.

12 April, Forêt de Crécy, Hauts-de-France. 50° 13' N. A tiger could hide in the shadows that beech trees cast over their own cast-off leaves.

They are the first shadows I have seen in three days. I haven't minded the rain, when it has come, but nothing suppresses spring like an overcast sky without the redemption of rain. Spring is defined by its sounds and its colours, even the colours of last autumn retouched by an April sun. I parked the car in a lay-by off the straight road that links Crécy with Forest-l'Abbaye via these woods. Within sixty seconds of entering the forest I heard an unfamiliar call coming from a nearby beech, high up above the point where the trunk divides into the inflorescence that is the crown. A downslurred, slightly crow-like call but softer and, comparatively at least, with a hint of sweetness. Immediately, a crow-sized, crow-coloured bird flew from behind a bough, landed on another a few feet to the left and shuffled sideways until it was hidden from view; so un-crow, so anti-crow. Then it flew from that tree for about 50 yards and landed, as far as I could tell, in the one I am looking at now, waiting for any movement, any sound that would confirm I have the right one. A crow-sized, crow-coloured non-crow is a black woodpecker.

The forest is ringing with nuthatch calls, a repeating, armour-piercing whistle, six or seven to a magazine, exchanged between lines of beech. In contrast, a stock dove's love song is quietly and erotically human; they are here, I suppose, as tenants of any vacated black woodpecker holes.

I am close to concluding that I have the wrong tree, when three deep, wood-chisel blows resound through the forest from high in the canopy at least 50 yards away. It is a commanding sound, like a call from another dimension, an upperworld order to us in the netherworld, magisterial and Wagnerian. There is no response from the nearby tree, so I walk around it, studying its vaulting for the briefest flick of black, and realise that the bird has eluded me. I walk deeper into the woods, but fail to find any further sign of the black-cloaked summoner.

In another part of the forest the light is scattered through the naked immature crowns of a stand of young beech. Bluebells, on the point of bursting open, shimmer a blue that is blanched by a wrapping of translucent cuticle. They appear like a mirage, a false frozen lake composed of shards of glacier. I make my way in a wide circle back to the car, and read the information board at the lay-by. There's a poster for an exhibition about 'Les Forêts Picards' that ran from 16 July to 29 September three years ago. There's an organigram of the hierarchy of the Office National des Forêts, and there's a large panel of information about this forest in particular. It explains, in detail, the composition (70 per cent beech, 30 per cent oak), productivity (15,000 cubic metres of timber and 15,000 steres of heating and pulping wood) and rotation (40 years) of the commercial operation. It goes on to set out the history (from AD 877), composition (roe dear and wild boar), letting arrangements (twelve-year leases) and timing (Saturdays, from the end of October to the end of February) of hunting in the forest. There is advice to 'users', which, among other things, forbids allowing dogs to roam free because of disturbance to game, and advises against entering areas where signs show that hunting is taking place. And, to quote my notebook: *fuck all about wildlife,* despite the forest being a Natura 2000 site, and therefore designated for wildlife conservation first.

Marquenterre, Hauts-de-France. 50° 15' N. Yesterday, Alexis and Christine said they intended coming here sometime during their week on the coast, reminding me that a minor diversion away from my route to Calais would bring me to one of the great estuaries of France – the Somme.

Twenty-eight years ago, when Jane and I came here on our honeymoon, we brought with us a coach-load of RSPB members from the Medway area. I had agreed to lead the group's day-trip across the Channel before we had set the date for our wedding, and it got us halfway to Paris for free.

At the entrance to the nature reserve, a banner announces the 'Festival de l'Oiseau et de la Nature'. The car park is crammed, and the café terrace full. I buy my ticket and set off along the nature trail to share the warm spring sunshine with a throng of visitors. At a lookout platform perched on a cliff formed of ancient dunes, I survey the Baie de Somme – the marshes, meadows and pools that stretch from the base of the cliff for a mile to the sea. I am surprised to hear the castanet-clattering of nesting storks, and scan the scene below me until my gaze alights on a distant black and white shape; their northward spread has reached the Channel coast since I was last here. I am even more surprised when a young woman in a zingy green jacket, Hedwige Letienne, informs me that the storks arrived back on 2 February (the day before San Blas Day), who knows from where. Hedwige tells me that she is an apprentice guide here, as she trains her telescope on the storks' nest for me to take a closer look. She did some of her training in Spain, and speaks Spanish more confidently than English; as I speak Spanish more confidently than French, we chat in our shared second language.

I take the longest trail, waymarked with red, which winds down the dunes, around the artificial ponds in the part of the reserve closest to the visitor centre and out into wilder, marshier country. It is warm enough for a common lizard to have found a spot to bask in, to take on fuel in its purest form – sunlight. Warm enough, too, for the air to rise and carry with it shimmers of dancing midges, and for these to pull in a score or so of Mediterranean gulls who hawk the insects. There is nothing more angelic than a back-lit Med gull against a cloudless sky. Its wings are pure white and in all senses immaculate. The muscled, inner part of the wing reflects light from below, off the water; the flight feathers and tail are translucent, seeming to emit a glow of their own, with their tips and edges more pellucid still, haloing the whole bird. Like angels, they have no visible source of nourishment, and their swallow-like pursuit of insubstantial prey is, to careless eyes, just some arcane, heavenly dance.

The trail leads through an avenue of pollarded willows and between spring-green water meadows, past the storks' nest I saw from afar. The synthetic-sounding song of lapwings stands out from the continuous rivulets of skylark song and the episodic accelerating trill of meadow pipits.

'Cattle egret!' I say out loud, surprising myself and a grey-haired woman who is standing nearby. 'Sorry, I was talking to myself,' I apologise in English, forgetting where I am, disoriented for a moment, like when you wake up in a hotel bed.

'It's OK,' she says with a smile and an accent that brings back childhood memories of Mireille Mathieu on TV, 'I do that all the time.'

The egret should not have been such a surprise. They have been breeding in Britain since 2008, but I still think of them as Mediterranean and African birds. It is surely climate change that has aided their most recent spread north, but they have an in-built propensity for expansion, facilitated by the growth in domestic cattle ranching. From tropical Africa they reached the Cape in 1908, and crossed the Atlantic in the 1930s, becoming established in South America within a decade or so. From there they colonised Florida in the 1950s and had reached Canada within another ten years. From Spain and Portugal they colonised the Camargue in 1958, and spread through France over the next half century.

The trail curves across the meadows in a wide arc that will eventually return me to the visitor centre. A brimstone butterfly; a waft of fox scent, like two-day-old spilt beer. There are sedge warblers everywhere – far too many to share the landscape for long, so most must be in transit. But they sing as if laying permanent claim to their few square feet. There is an audit of incoming migrants to be made – sedge warblers and blackcaps in abundance, willow warblers scarce, a few whitethroats and lesser whitethroats, no cuckoos but a briefly heard nightingale. As I approach the end of the trail, a spoonbill appears from over a stand of tall pines on my left and glides over me, to land in a field about 100 yards away. Another follows soon after and lands in the same place, where I notice, through gaps in the surrounding vegetation, that there are several more. One takes off and returns in the opposite direction, carrying a stick. I stand for a while and watch as the traffic increases, so that at any moment there is at least one bird crossing over, either gliding down to the right or flapping up and to the left, with a stick or a beakful of grass. A spur off the main track leads in the direction of the colony to what I take to be a large hide, but when I enter, it is in the form of a grandstand, open-fronted with two tiers of seating, facing across a clearing in the pines to the colony. I take my seat, along with about thirty other people. Fifty yards away, the spoonbills are dropping into the tops of the trees with their sticks, while others are stealing from their neighbours, which include white storks, grey herons, little egrets and a pair of cattle egrets. Another young woman in the green uniform of the reserve, Cécille Leroux, explains that the growling, gargling noise we hear is the sound a male little egret makes when trying to seduce a female.

I don't remember a heronry being here in 1988, and feel sure there were no spoonbills back then. Cécille tells me that the spoonbills first nested in 2000, that there are forty pairs at the moment, but another twenty are expected to arrive by the end of April. There are twenty pairs of storks, she says, three of them sedentary, originating from reintroduction schemes in France, and seventeen pairs that migrate here each year. Cécille sets up her telescope for an English couple to use, and explains in English that the young herons they can see are two weeks old. I spot a night heron, typically almost totally obscured in the trees. I am about to point it out to Cécille, when its mate arrives, carrying a stick, and flies across the view to a noisy welcome at the nest.

A middle-aged woman has some photos of birds she has seen on her walk round the reserve, and she asks Cécille's help to identify them. A willow warbler – *pouillot fitis* – that has just arrived from Africa, news which brings an audible gasp from the woman.

'*Pouillot fitis.*' She whispers the name with utmost reverence, and swipes open another picture. A little grebe, says Cécille, the smallest grebe.

'*Oh! Grèbe castagneux!*' whispers the woman. '*C'est super-jolie.*'

Wissant, Hauts-de-France. 50° 53' N. Every evening for the past five nights, the report from home has been the same: the swallows have not yet returned to The Hamlet. Each time, I have told Jane of the swallows I have seen on this side of the Channel, a day's flight from Yorkshire. This evening I saw a thousand or more, heading north along the coast. I could see Kent, the county where I grew up, in the background. The swallows will be settling in some reed bed to the north of here, and will be in Denmark or Germany tomorrow. A few may cross the Channel into England. High above the swallows and 2 miles or so out into the Channel was a skein of geese, probably brent geese on the first day of their emigration from the estuaries of the Channel coast towards the high Arctic tundra of Spitsbergen or Russia. Yang Lian, the Swiss-born, Chinese-raised, Hackney-based poet, sees the skein as a symbol of exile and longing; and notes that the shape of a skein of geese is the Chinese character for 'person'.[7]

The geese will reach England by nightfall, and tomorrow I shall follow.

United Kingdom

14 April, South Downs, Hampshire. 50° 59' N. Jane drove down from Yorkshire yesterday and met me at Petersfield with news from The Hamlet: swallows, none; but in the old barn behind the house, the pied wagtails are back and nesting under a loose ridge stone, as they have every year since we moved there; a pair of stock doves are in residence in the same building. It will be good to get back there tomorrow.

This morning we set out from the Hampshire village of East Meon to walk a few miles of the South Downs Way. At a sign announcing the hamlet of Coombe Cross we turned left off the lane and uphill along a chalk path through the woods. At the base of the sign someone has planted tulips and Spanish bluebells, and as we climbed the hill, we noted a few clumps of native bluebells along the path, among the primroses and wood anemones. The flowers were still locked away in short, unopened spikes, and Jane pointed out one clump with the robust stature and side-to-side arrangement of flowers that suggests impure stock. It is one of the quiet tragedies of English woodland that its greatest glory is slowly fading. The native bluebell, *Hyacinthoides non-scripta*, is not unique to Britain, but the shimmering blue carpets that create surreal lacustrine visions and waft air-cleansing fragrance, are. It is a unique blue, the colour of late afternoon, and a unique shimmer, as if the delicate, crook-shaped stem were designed to catch any ground-level drift of air. *Hyacinthoides hispanica,* the Spanish bluebell, probably first appeared in Britain's gardens in the seventeenth century. It wasn't until 1908 that it was first noticed in the wild, and then more than half a century before the first hybrid was discovered in 1963. *Hispanica* can swamp *non-scripta* genetically, producing a hybrid that itself produces highly fertile seed, so that botanists refer to it as a hybrid species. Thus the hybrid is effectively a new, man-made plant, more fertile than the native bluebell, and gradually replacing it with versions that lack its gracility, shape, depth of colour, fragrance and right of tenure.

The green lane is of tackily moist chalk, and dog violet, dog's mercury and dog-crap bags grow at its edge. We stop at a contorted beech of ten boughs, each the size of a small tree, and look over the valley between here and Small Down, at a white slough of exposed chalk in the steep grassland opposite, like a potter's thumbprint. I hear a distant willow warbler; they are late.

We pass along the high fold in the Cretaceous chalk known as the Winchester-East Meon Anticline. Fields have been ploughed and harrowed: from a distance, they have the colour and granular texture of stone-ground wholemeal; up close their surfaces are 50 per cent chalk stones and flint. At the summit of Old Winchester Hill we stand in the centre of a flat, oval field of about 10 acres. From this plateau we look across the Downs to the Solent, and across that to the Isle of Wight. A pair of buzzards hang in the view, and seem as fixed into the scene as the distant skyline of Chichester, 18 miles away and 648 feet below. For a century they were a rare bird here, infrequent visitors from among the survivors in the New Forest, wiped out everywhere else in the county. By 1905 the buzzard was

> a very scarce resident in Hampshire and seldom allowed to rear its young, but occurring in various parts of the county in autumn and winter.[1]

In 1778 it was apparently familiar to Gilbert White, whose entire seventy-three years were spent in southern England, all but about ten in these Downs. On 7 August that year, in a letter to the lawyer and naturalist Daines Barrington, he wrote:

> A good ornithologist should be able to distinguish birds by their air as well as by their colours and shape; on the ground

> as well as on the wing, and in the bush as well as in the hand. For, though it must not be said that every species of bird has a manner peculiar to itself, yet there is somewhat in most genera at least, that at first sight discriminates them, and enables a judicious observer to pronounce upon them with some certainty. Put a bird in motion
>
> *et vera incessu patuit.**
>
> Thus kites and buzzards sail round in circles with wings expanded and motionless; and it is from their gliding manner that the former are still called in the north of England gleads, from the Saxon verb *glidan*, to glide.[2]

That name died out when the red kite was lost from the air above northern England in around 1870, although it survives in place names within a few miles of The Hamlet: Glead Hill, Gled Hill and Gled Hall. They are back, of course, to symbolise changed attitudes and conservation success. There are over 100 pairs in Yorkshire this year, and about 150 pairs in northern England as a whole, less than two decades after their reintroduction. The old name did not return with them.

We saw one this morning. In a field above East Meon a sheep lay dead and browned with dried blood. Six crows attended, and the kite arrived along the anticline from the west, slowly surfing the updraft as we passed along the road above. Its fish-tail flexed in response to minute irregularities in the air flow; its wings beat languidly only when they had to, such as when the bird suddenly decided not to drop in on the feast but to return whence it had appeared. The red kite was exterminated from Hampshire when the last recorded nest was taken down in 1864.[3] The next record for Hampshire was in 1956, here at Old Winchester Hill. It

* And its true self is revealed by its movement (paraphrasing Virgil).

finally returned to breed in 2003, having followed a chalk line from the Chilterns, where its reintroduction to Britain began in 1989.

In 1974 there would have been few buzzards – and no kites – to hang and glide about the escarpment. In that year the Old Winchester Hill panorama was etched into stainless steel, and a panel fixed in position at the highest point, in the centre of this field. It is still there, a picture-map in egocentric projection, guiding the stationary observer across a field of near and distant landmarks:

Chichester Harbour 18 miles S.E.
Portsmouth 12 miles due S.
Isle of Wight 18 miles S.S.W.
Calshot Power Station 16 miles S.W.
Stocks Farm 2 miles S.W.
Southampton 14 miles W.S.W.
Beacon Hill 2½ miles W.N.W.
Privett church 4½ miles N.N.E.
Beacon Hill 25 miles N.E.

Place names are our means of transport across the centuries, always with the caveat that they say as much about the namers as they do the places. They are the pins that fix the overlays of landscape history. Chichester (or Cesseceastre) was the Romano-British city of Noviomagus Reginorum before it was renamed in AD 477 by its Anglo-Saxon captor Ælle, after his son Cissa. To read the landscape of the time before the Romans, one must go downhill to the coastal plain and look up. Two and a half thousand years ago this hill would have shone against the north sky. By day the bare white chalk of the Iron Age defensive bank would serve as the mark of some Celtic chieftain's power, and on moonlit nights its glow would hover over fields and forest to the howling of wolves.

A thousand years before that, the long, broad crest of the hill with a lost Bronze Age name would have been newly denuded of trees, and the forest cover on the other hills and in the lowlands would be patchy. There would be places where the view to the summit was uninterrupted – important places, from which a dozen or more white chalk mounds, their faces weeded and cleaned every year, could be seen. The people in the scattered communities on the slopes and flatlands would look up and remember the important people buried under them.

Today, it is a landscape of skylark-ringing barley fields and grass leys. George Meredith was born within this view, in Portsmouth, and fell under the spell of the bird that 'drops the silver chain of sound' and inspired 'The Lark Ascending'. It was 'a pinch of unseen, unguarded dust / The dust of the lark that Shelley heard,' in Thomas Hardy's poem about a poem. All three men lived in the chalk downland 'where ripple ripple overcurls / And eddy into eddy whirls': words which Ralph Vaughan Williams heard and, it seems to me, set the poem, not the bird.[4]

I wonder how much this view has changed since Gilbert White's day. The temptation is always to paint more trees into an imagined past landscape, whereas here in the South Downs there is a greater acreage of woodland today than for many centuries past. The watercolours of Copley Fielding (1787–1855) depict the South Downs of the day. Whether looking across their summits and down over the plains below or looking up at them from the lowlands, he invariably showed an upland landscape utterly bare of trees and shrubs. The Downs were known as sheep-walks: originally cleared for agriculture in the Neolithic and Bronze Age, it was the wool boom that peaked in the fifteenth and sixteenth centuries which created the flower- and butterfly-rich chalk grasslands whose fragmentary relics are protected today. The

Old Winchester Hill National Nature Reserve will come into its own two or three months from now, when the greater-butterfly, bee, frog, fly, common spotted and fragrant orchids bloom and the chalkhill blues and dark-green fritillaries fly. Jane and I are content to find a less ostentatious beauty lying among the shorn grass – the hairy violet *Viola hirta*, a widespread but declining chalk-land species, and the preferred food of the caterpillars of the dark-green fritillary.

Twenty Herdwick sheep – a hardy, blue-grey Lakeland breed – are grazing the north face of the hill fort. We walk among them, skirting round a steep combe that is part grassland, part mixed woodland. A single swallow circumnavigates the combe in an erratic-purposeful pattern. Our path curves round its southern rim, a crescent-shaped fold in the hill that falls steeply to our left, where a male wheatear flits between ant hummocks. Jane spots a movement on the slope a hundred yards ahead of us and freezes. A pair of green woodpeckers digging for ants. The male is to the right of the female, up-slope, and burnished green, gold-green, outshining his grass-green mate. A second male calls from the wood on the opposite tine of the crescent, and our male stops digging for a moment. The other male appears from the trees, and flies down the combe, parallel to our path, as if to show himself to the ant-hunting pair. From the wood at the bottom of the valley, he calls again, a loud, ringing laugh or 'yaffle' that echoes around us.

> April 14, 1771, Selborne. Swallow appears as last year amidst frost & snow!

Between 1768 and his death in 1793, Gilbert White kept a diary, his *Naturalist's Journal*, which was compiled posthumously for the second edition of *The Natural History and Antiquities of Selborne*. Jane and I have never been to

Selborne, and as it is only 12 miles from Old Winchester Hill, we have come to The Wakes, the house White grew up in and returned to when he inherited it from his father in 1758. When White began his extensive correspondence with two of the leading gentleman naturalists of the age, Thomas Pennant (1726–1798) of Flintshire and Daines Barrington (1727–1800) of London, he was consciously recording his enquiries and observations for posterity. Between 1767 and 1787 he wrote thirty-five letters to Pennant and sixty-six to Barrington. These, along with another nine unsent letters that he wrote to serve as prologues, make up *The Natural History and Antiquities of Selborne*.

We find White's original manuscript displayed behind glass in an upstairs room, open at a letter to Barrington dated 19 March 1772. As I read it, I realise that it has been selected for display because it contains a reference to today's date. The previous 4 November, the letter explains, White and a friend were surprised to see three house martins on the coast at Newhaven.

> I am more & more induced to believe that many of the swallow kind do not depart from this island; but lay themselves up in holes & caverns; & do, insect-like and bat-like, come forth at mild times, & then retire again to their *latebræ*. … And I am more of this opinion from what I have remarked during some of our late springs, that though some swallows did make their appearance about the usual time, viz., the thirteenth or fourteenth of April, yet meeting with an harsh reception, & blustering cold N+E: winds, they immediately withdrew, absconding for several days, till the weather gave them better encouragement.

White, Barrington and Pennant were protagonists in a debate that was at least 2,000 years old, since Aristotle wrote in the *History of Animals* in 350 BC that many birds do not migrate but go into hiding. Aristotle appears to have based this view on reports of swallows being found alive but without feathers in holes during the winter. In the course of the correspondence, White is revealed as a firm believer in migration, unlike Barrington, but not a complete disbeliever in the possibility that some birds may hibernate as well. But he refused to go on hearsay, and went to great lengths to find hibernating swallows and martins, all of which are set out in the letters and journals. His *Naturalist's Journal* tells us, for example, that on 5 April 1781 he

> searched the S:E. end of the hanger for house-martins, but without any success, tho' many young men assisted. They examined the beechen-shrubs, & holes in the steep hanger.

The previous 14 October he had seen house martins going to roost in the spot, and resolved to leave the area undisturbed until early the following spring. The date chosen for the examination is characteristically thoughtful: close enough to a typical arrival date for them to complete very nearly a full term of hibernation, but early enough to ensure he had not missed them.

The journal and the correspondence shows White growing increasingly convinced of the normality of migration, and failing to find convincing evidence of hibernation year after year, although he never came fully off the fence on the question. Since then, two centuries of ringing and, lately, satellite studies have shone ever more light on birds' extraordinary movements. Aristotle's belief in featherless birds spending the winter in clefts in the rock is consigned to the realm of fantasy. But at the beginning of this year a discovery was announced that may explain the origin of that view. On the barren Moroccan island of Mogador, Eleonora's falcons have been found to imprison small birds by trapping them alive. The falcons are specialists: they breed on Mediterranean islands, waiting until late in the autumn to raise their young on the enormous numbers of southbound migrants that use the islands as stepping-stones for the long sea crossing. During a census of the falcons in 2014, Abdeljebbar Qninba of Mohammed V University in Rabat, Morocco and his colleagues came across small birds trapped in deep cavities, their flight and tail feathers removed. The migrants tend to arrive in waves, often overnight, or pulled in by bad weather. Crippling and imprisoning prey might be a means of dealing with the peaks and troughs of abundance, keeping fresh food nearby for the falcons to feed to their young during quiet periods. At the end of the season, the falcons undertake their own long journey over Africa to winter in Madagascar.

20 April, Wicken Fen, Cambridgeshire. 52° 18' N. On the fen immediately east of Ely railway station is a remnant of the Ouse floodplain, bounded by the city and railway on one side and, I suppose, intensive farming beyond the curve of its far edge. A grid of drainage dykes, shallowed by decades of silt and leaf-fall, is marked by parades of old willows. Some of them have fallen, decades ago, and lines of upright branches have sprouted from their prostrate trunks. Now each branch is a tree in its own right, almost as thick as its clone-parent. For each fallen original, there are two, three, even four new trees in perfect alignment. Should one of these fall, as it surely will, it too will create a new troop, clones of itself, marching downwind, one pace every few decades. Their name *Salix* is from the Latin *salire*, to leap, and there must have been places in old Fenland where a fair mileage was traversed by slow armies of wind-thrown willows. The Britannic word for willow – *helig* – is the origin of the city's Saxon name, and with an easy slip of two tongues, Willow Island became the Island of Eels.

I have come a few miles south-east of Ely, to Wicken, drawn in a way by a convergence of ley lines through space-time. Looking west, the Brecks of Suffolk and Norfolk stretch beyond the horizon behind me; ahead lie the Fens of Cambridgeshire, Norfolk and Lincolnshire. Etched somewhere upon this landscape is the route taken by Robert Rose's son John, who will have passed this way one day, around 1840, aged about twenty-one. In the land he left behind, the wind denuded the Breckland soils, as it always had; the newly planted windbreaks would take decades to grow into full defensive array. To the west, however, new steam pumps had opened open up thousands of Fenland acres to year-round farming on the finest soils in the country. In places the old fens remained, where wind pumps had not given way to steam; in the case of Wicken and Burwell Fens, they had never been effectively drained, remaining to this day. The fens that John Rose crossed to reach Tydd St Giles would have included some of the richest farmland, and the wildest marsh, in England. John married a Tydd girl, Mary Bliss, but two generations before, the Blisses themselves had migrated from the harsh northern Brecks.

I am here to meet Ralph Sargeant at the suggestion of my friend Joan Childs, who for years was an RSPB colleague and since 2014 has been the National Trust's Wicken Fen manager.* The path to the reserve's reception building crosses a pond, and leaning on the handrail is a man in his seventies, stocky and slightly portly, with shoulder-length grey hair and a Charles Darwin beard. He is wearing a checked shirt and green sleeveless fleece, and watching the water. I lean on the rail a foot or so to his right, and look down at the pond too.

'You'll be Ralph' I say, and we shake hands. I garble an explanation for wanting to meet him, aware as I speak that it doesn't make sense, probably because I don't really know the

* Joan left the National Trust later in 2016 to join the North York Moors National Park Authority.

reason myself. But I mention my family connection to the Fens, and we find common ground.

'My Dad was from the Fens,' I tell him. 'From the Wisbech end. Some of my ancestors were from Wicken, though. Well, mainly Soham. My great-great-great-great-grandmother was from Wicken.'

'What were their names?'

'The Soham ones were Reeve, and my ancestor Samuel Reeve married Martha Wright from Wicken in about 1818.'

'Well I don't think there are any Wrights in Wicken now, but I know plenty of Reeves in Soham.'

Lily Reeve was born in 1898 in March, whose name may betoken its marshland setting, but her immediate forebears had come from undrained Wicken and wind-pumped Soham, arriving in the Isle of Ely as recently as the 1870s. Lily married Arthur Rose, who was the grandson of John, and my grandfather.

'Shall we walk and talk?' I suggest.

'Happy to talk, but not walk. Bin a bit poorly last few weeks.'

We sit at a picnic bench outside the National Trust café, where Ralph has chosen Earl Grey tea and a cheese scone, and I have copied him. I tell Ralph that, as I was researching my family history, I came across a John Sargeant and family in the 1911 census, in Chapel Lane.

'That's right, John was my grandfather. He was a builder. He built two houses on Chapel Lane for £26, sold one and married my grandmother when he had money. He built the Mission Hall and almshouses. They were only pulled down two years ago.'

'And there are two sons, James, born 1895, and Baden, born 1900.'

'Baden Powell Sargeant, he was my father. He was a French polisher, till he got TB. Then he was a builder.'

'Joan tells me you've been a bit of a fixture at the Fen for years.'

'Yeah, she told me you wanted to meet an old-timer. I started working here officially in 1971 or 1972. Colonel Mitchell, the Head Warden, sent for me.'

'Sent for you?'

'Yeah. Well, I was always here anyway. I had a building job, and helped out on the local pig farm till Unwins took it over. Weekends I'd be here pike fishing and catching daphnia for my tropical fish. I used to catch things for the Colonel; there was a rare spider he was interested in.'

'Do people still catch pike?'

'Only to put 'em back. They don't know how to prepare it. You need to soak 'em in salt water for two days, like cod. They don't know how to get the scales off, so they fly everywhere and stick to the ceiling.'

'So you came to work here for the Colonel.'

'Yeah, him and Wilf Barnes, the keeper. I got £21 a week. January 1972 it was, now I think about it. My first job was cutting reeds, and cleaning them ready for roofing. And cutting pathways with an old Mayfield offset-blade cutter. Then July I'd be onto the sedge – cutting it for the thatchers. They use it for ridging; each thatcher had his signature ridge style. I think a lot of it comes from Poland now. We used to clean the sedge but they didn't like us doing that, they said it didn't pack down as well.'

'What do you mean, clean it?'

'Getting rid of the seed-heads and old leaves. We'd sell all that litter to the potato growers for layering. One winter a male bearded tit used to come and feed on the cut reed seed-heads while we were working. Another year there was a Cetti's warbler with us the whole time we worked, the first one. That year a pair had two clutches on Harrison's Drove and within three or four years they'd ringed sixty of 'em.'

I remember being taken to see my first Cetti's, at Stodmarsh in Kent in 1972, the year they first colonised England. By 2008 their British population reached 2,250 singing males.

I have been reading James Wentworth-Day's *A History of the Fens*, written in 1953. Wentworth-Day's family lived in The Marsh House, Exning for 300 years, within hearing distance of Wicken Fen. He was a fanatical wildfowler and loved the old Fen practice of punt-gunning. He bought Adventurers' Fen, which the artist and writer E.A.R. Ennion depicted beautifully in his book of the same name. Adventurers' was drained by the War Agricultural Executive Committee in 1941, for which Wentworth-Day writes an eloquent grudge. He became a minor political figure and major embarrassment to the Conservatives, making statements on race relations that would be unpublishable today.

'You don't want to believe everything in that book,' says Ralph, 'but I remember some of the old characters he writes about. Bert Bailey, "Soldier" Bailey, he was the last peat-digger – on St Edmund's Fen. He was in the Maid's Head every night. Eventually the brewery let him have a free pint a day.'

We speak for an hour, then I get up to buy us some cake and more tea, and to give Ralph some respite. His memories and opinions have flowed like the Cam, but I sense him tiring. He had a triple heart bypass in 1999, he tells me, and spent thirteen weeks easing himself back to work. He retired in 2007 but continues to help out with bird recording and showing visitors around.

'I mustn't take up too much more of your time, Ralph', I say when I return.

'No bother. I was just thinking about the last coypu we had here. I caught it by the tail but couldn't hold onto it. We got it in the end. It weighed 28 pounds. I buried it in the garden and my French beans grew a foot higher there.' Then, without a pause, and as if to draw our conversation to a close, 'I had a hip replacement in July, and have been in and out of hospital ever since. Leaky heart valve apparently. I'm back there Thursday.'

'You look after yourself, Ralph, it's been good to meet you.'
'Yeah, it's been good to talk about them old times.'

At the edge of Sedge Fen I peer into water the colour of Ceylon tea, my eye diverted by the flash and zip of whirligig beetles. They reflect sunlight with a purple-pewter lustre, and somehow concentrate it as if they were the light source. The water is like a mirror, and to look down into it is to look up through the reeds and through the leafless ash and catkinned willow trees, to a sepia-tinted blue. I try to fathom the beetles' motion. Alone, *Gyrinus substriatus* does not so much gyrate as patrol in jerky linear movements. In twos, they pirouette and pursue, slow-slow, quick-quick, slow. An encounter involving any greater number leads to a complex spinning and weaving, to an unintelligible algorithm that reminds me of fairground dodgems.

It is the archetype of a spring day: bright, with well-delineated shadows; soft, warm air scraped at by sedge warbler song and flickered by yellow brimstone butterflies; the first mild strike of pollen on the eyes. In the white-straw hummocks of old-growth sedge, new glaucous spears show through, and there are patches of yellow flag leaves. I follow reserve trails that trace the old lodes and droves, whose names are a roll-call of past figures in a landscape. Some are anonymous collectives, like the commoners who had the right to dig peat at Wicken Poor's Fen. There were the Adventurers who invested – unsuccessfully – in draining the fen that Wentworth-Day bought and lost to the Commissioners, themselves commemorated in Commissioners' Drain.

I cross Monk's Lode on a flat wooden bridge and turn left and then right along Moore's Drove. But I return to the lode when I hear a cuckoo – my first in Britain this year – calling from some tall willows 200 yards or so along it to the east. As I walk towards the trees, waiting for the cuckoo to call again, I look into the clear water of the lode. Water lily leaves have begun their rise towards the surface, their edges furled inwards like scrolls. The cuckoo calls again, from deep inside the spinney, and I return to the trail, turning right onto Harrison's Drove, alongside Adventurers' Fen. Now part of the National Trust land, Adventurers' is being restored to wetland, a small step towards a new vision for these great levels. Four mud-caked konik ponies are neck-rubbing a gate. They have the coloration of seal point Siamese cats, and kohled eyes like pharaohs.

Alongside Wicken Lode, to the descending semi-quavers of willow warblers, I catch a scent that stops me as though snagged by a bramble. I track it down to the catkins of sallow, a hyacinthine perfume that, like that of the primrose, is waft-dependent.

Purls Bridge, Cambridgeshire. 52° 27' N. The full moon rose three hours before sunset, and now that the sun is down and its twilight extinguished, the water meadows of the Ouse Washes are streaked with white reflected sunlight, a blaze that stretches directly before me like a comet's tail. There are other lights at the eastern horizon, specks of street light and window light from Ely 6 miles away. There is a strange light too, hovering above the city like a UFO. Through binoculars, if anything the cathedral's octagonal Lantern Tower, 171 feet above the ground, seems even more extra-terrestrial.

I have come to sit and listen to the night, and then to sleep with the sounds and wake with the sun and the new shift of sound-makers. When I am ready to get into my sleeping bag, the moonlight will guarantee me a night at the margins of sleep, of heard, dreamt and half-dreamt sounds. When the sun's twilight still coloured the sky, I could scan the washes through binoculars and watch the traffic of birds arriving to roost. In the distance, there is a flock of black-tailed godwits, *Dei ingenium*, as Isaac Casaubon unwittingly punned.[5] They represent continuity: in 1611 Casaubon noted that they could be bought at Wisbech, having been fattened in captivity, for five or six halfpence, and that they were 'wonderfully commended'.

Something disturbs the godwits, and they take to the air in the last tinted moments of twilight, red and dark, like dried blood, rather than their brick-red daylight colour. Their English name is derived from their call: two notes, the second higher than the first, with a grace note joining the two pitches and an interrogatory inflection – 'to-t'whit?' But from a flying flock all calling, there is an aleatory effect that for some reason I imagine as the sound of trickling water played backwards. The twilight's dimming coincides with the moon rising and brightening, causing a shift from red-tinted chromaticism to silvery monochrome. The light is now more directional, from the east-risen moon, rather than from the sky itself. The water can be heard: swishing when unseen ducks fly in and settle; slapping at the broad feet of mute swans, and accompanying the windmill-sail swipe of their wings. The long, pure piping of a redshank is a sound designed for distance, for telecommunication. Another replies, its call resolving into a wheeling yodel. The modest chirrup of teal, the lime-sharp stab of coots. And then, the sound I have most been hoping to hear.

I remember when I first heard it, close to here, when I was about thirteen on a visit to my grandparents. I watched as a strange, long-billed bird rose only to plummet back into the field with a sound like a balsa-wood plane. I struggled to identify it from the descriptions in my *Observer's Book of Birds*. I could not reconcile the pictures of the most likely candidates with the descriptions of their calls; no sound ascribed to them came close to that flat airborne bleat. Finally, I realised that this bleating was no cry from within, but a rush of air through a snipe's tail feathers, a specially adapted wind instrument that rose from the marshes to deliver its vibrations across a wide open sky. On subsequent visits, when I was a little older and unaccompanied, I would walk alongside these fields after dark, listening to the night sounds, wild swans and redwings in winter; and in spring, a dive-bombing snipe, leaving its unique feather-music hanging in the clean air.

I would usually spend the evening at The Ship, the front room of Will Kent's house in Purls Bridge, where I would sip my pint, write my notes and watch the same three men play dominoes. Will Kent would sip tea, and few words would cross the room between any of us. I never heard a pint ordered: now and then Will would rise from his chair, a slow, grey hulk, collect a glass or two and disappear from the room, returning with fresh beer. A hundred years ago it was Will's father who kept The Ship. He was descended from Fen Tiger stock, a breed of men who in recent memory lived off ancient wetland skills alone. Dr Katherine Heanley came here with her friend Gertrude and kept a diary.

> *September 16th 1915.* And now we are at Purles Bridge, which is no bridge but two square public houses,* one on each side of a secondary road which leads from Manea to the Old Bedford River. One Inn is called the 'Ship' and the other 'The Chequers'. Excepting that the 'Ship' has a seat outside and the 'Chequers' has not they are both alike in all respects. Mr. Kent keeps the 'Ship' and we are going to have lunch on his bench before we leave this enchanting spot, on the bank of the Old Bedford River; cut in or rather begun in 1630+? by the Russell Company; 21 miles long in the finish which was I know not when. Here it is after 300 years, draining the land, feeding the ducks of Mr. Kent and giving drink to a troop of young calves, who drink deli-

* Neither exists today.

> cately and make no snuffling noise. It runs past the E. end of Vermudens* Drain from the Ouse at Earith to the Ouse at Denvers sluice. We lost the train again this morning to Manea, the chambermaid's fault this time, who told the bus driver the wrong time to fetch us, nearly an hour after the train had gone; and so we had a very pleasant motor drive here which cost 10/-.

This Will Kent was the last man in the area to shoot a punt gun for a living. He also grew willow rods from the osiers that still grow here; he will have cut them, seasoned them, peeled them and woven them into eel traps – long *grigs* and short *hives*. He will have wielded his *glaive* at the numberless eels and his *dart* at the great river pike and aimed his punt gun at flocks of mallard, teal and coot.

The literacy rate among earlier Fen Tigers would have been little more than zero, leaving few first-hand accounts of their extraordinary lifestyle. Where, or rather how, William Hall was educated is anyone's guess. He was born in 1748 on Willow Booth, a tiny fen island near Heckington, then the remotest part of the Lincolnshire fens. Somehow he learnt to read, and this he did with a passion, and to write, with equal dedication. He styled himself Fen-Bill Hall, left a richness of prose and poems describing his life as a fen-man, and went on to own an 'antiquarian bookstall' in King's Lynn, where he died in 1825.

He begs his readers' indulgence with regard to his imperfect education, having lived 5 miles from the nearest school, and having had no more than six months' schooling:

> Where Ducks by scores travers'd the Fens
> Coots, Didappers, Rails, Water-hens,
> Combined with eggs, to charge our pot.
> Two furlongs circle round the spot.

* *Sic* Cornelius Vermuyden (1590–1677).

Fowl, fish, all kinds the table grac'd,
All caught within the self-same space;
As time revolved, in season fed,
The surplus found us salt and bread;
Your humble servant, now your pen-man,
Liv'd thus, a simple, full-bred Fen-man.[6]

One detail from Fen-Bill's output brings back from across four decades a memory of a magical spring evening here at Purls Bridge. I had walked from Manea village across the old pontoon to the RSPB reserve at Welches Dam, from there back along the Old Bedford River, and on to Welney. I returned at the end of the day the way I came, light fading. Never wanting to waste a scintilla of coppery fen twilight, I slowed my pace as I approached The Ship. On the opposite side of the river, in a black, ploughed field, ten ruffs were jousting, their false and exaggerated belligerence dangerous but not intended to harm, but rather to display and compare finery, in their own version of chivalric combat. The bronzy ruff of one in particular stood out in the fading light and against the soot-black soil, before they suddenly flew off, perhaps when one took his eyes off the game and noticed me.

Fen-Bill Hall caught them in *springes,* nets fitted with springs and set where the birds were expected to congregate on their lekking grounds. They were caught alive, kept in captivity and fattened on milk. The Reverend James Francis Dimock, contributing to Yarrell's monumental *A History of British Birds*, recorded that reeves (female ruffs) bred in Cawlish Wash near Spalding, and that ten dozen birds were fattened and sent to Leadenhall market in 1824.[7]

At about 11 p.m., I dig feet-first into my sleeping bag, and at the same moment a call from across the moon-silvered meadows seals the day. A crane, the bird that most symbolises the great natural wealth lost to the drainage of the Fens, lets fly a single, long, loud and multiphonic signal, like the end of a shift. Absent from the Fens and from the UK since the 1600s, cranes have returned to eastern England and their numbers are slowly climbing. This year there are a record forty-eight pairs in the UK, with nine pairs in the Fens.

> There stalks the stately Crane, as though he march'd in warre,
> By him that hath the Herne, which (by the Fishy Carre)
> Can fetch with their long necks, out of the Rush and Reed,
> Snigs, Fry, and yellow Frogs, whereon they often feed:
> And under them again, (that water never take,
> But by some ditches side, or little shallow Lake
> Lye dabbling night and day) the palatt-pleasing Snite,
> The Bidcocke, and like them the Redshanke, that delight
> Together still to be, in some small Reedy bed,
> In which these little Fowles in Summer time were bred.

So wrote Michael Drayton in *Poly-Olbion*, a 15,000-line topographical and cultural description of England and Wales written in 1612. About sixty lines are devoted to a list of fenland birds:

> The Buzzing Bitter sits, which through his hollow Bill,
> A sudden bellowing sends, which many times doth fill
> The neighbouring Marsh with noyse, as though a Bull did roare;
> But scarcely have I yet recited half my store:
> And with my wondrous flocks of Wild-geese come I then,
> Which look as though alone they peopled all the Fen.

Drayton describes a landscape that was once mostly water, inhabited by an amphibious people who lived on low islands hardly discernible today. Village names tell of a scattered and causewayed archipelago: Manea, Welney, Ely, Stonea and

Shippea, with their Saxon suffixes. Their church spires, overlorded by the great octagonal tower of Ely, still punctuate a skyscape shared with grain silos and serried pylons. From the air, at certain times of year, silted channels show through the peat. They are true fossils: once-wandering watercourses that redrew new filigree at the end of every winter, now indelibly the signature of the year in which each field was drained. Tiled over this ghost-map are wide fields delineated by arrow-straight waterways that zigzag the roads.

22 April, Rhostryfan, Gwynedd. 53° 06' N. I hear that yesterday a new red kite nest was discovered 140 miles north-east of here in Eccup near Leeds. Hanging in the tree next to it was a corpse: a dead kite. An X-ray has revealed a fully formed egg in its belly, ready to be laid, along with several shotgun pellets. It was probably shot on the nest while it was incubating its other three eggs. The red kite has, over the last twenty years, ceased to be the bird – *the* bird – of Wales. Nor should it ever have been. It became Welsh the way the Welsh language did, when Wales became the place from which they were never extinguished. Now the kite has returned to England, to thrive and to suffer.

I have a mug of tea by the Afon Wyled, a stream that runs under the road and alongside the back of the house where Siôn, Non and new-born Ethni Dafis live. Its retaining walls are a natural fern garden with splash-watered common polypody and hart's tongue draped over the rough-cut stones. The mug is printed with the slogan 'Listen to the World'. I am. A mountain stream, contained between stone revetments, channelled through a stone village, resonating off chapel walls; and jackdaws, making echoes for fun. There is something of the smithy about it, that combination of murmur and hammered anvil. The jackdaw, accompanied by water, *is* the bird of Wales. Today it is, anyway.

Before Siôn and I set off through the village I look to the left, north-west, along the road I have just driven up from the coast, and realise for the first time how high we are above sea level.

'That's Dinas Dinlle,' says Siôn, pointing out the thin wedge of coastal plain that is visible between the houses, 'we'll get a better view in a minute.'

We turn onto a disused railway track that follows a contour line on a hillside falling steeply to our left. The track takes us through open woodland of oak and old hazel coppice, and a stream passes under us through a culvert built when the quarry line was created to ferry slate to Dinas for transhipment to the London and North Western Railway. I notice a small clump of marsh marigolds, like huge buttercups, at the edge of the stream, and walk on. They are a typical early spring flower of boggy places, and they have been at their best at The Hamlet for the past week. I realise Siôn has stopped to admire them.

'They weren't open yesterday,' he says. 'You can't match that yellow.'

It is a few years since we have seen each other. I spent 2010 seconded as acting Director, RSPB Cymru, based mainly in Cardiff but with a North Wales office in Bangor, which I visited as often as I could. Siôn worked as an ecologist for the RSPB in Snowdonia, surveying farmland birds and working with farmers on their agri-environment schemes. His was fieldwork, and on the few occasions we were together in the Bangor office, I looked forward to his reports from the front line, delivered with passion, eloquence and an accent designed for both. We didn't know each other well, but something he once said has stuck in my memory: 'Conservation organisations don't speak to farmers about nature the way farmers speak to each other about it, in English or in Welsh.'

Spring has been creeping north through Britain and up through the foothills of Snowdonia, as it has through the Pennine foothills at home. We cross another stream, Afon Carrog, which tumbles along a scar cut by its own force into the hillside coppice. On its banks are two more early spring flowers, golden saxifrage in the splash zone, and one Siôn calls windflower. I had almost forgotten this other name for the wood anemone.

'The local name for this spot is Bicall,' says Siôn. 'I have a hunch it's derived from the North Walian word for spear,* with all this hazel coppice. Have you heard of Kate Roberts?'

'Yes. No. I think I have.'

'A lot of her work is set here. Right *here*, in fact. I never liked her stuff much when I was at school. Everyone learns Kate Roberts† in the Welsh-medium schools. We're taught she's *Brenhines ein llên* – Queen of our Literature. I just thought it was all dry and depressing. She was from Rhosgadfan, which is just over there, the next village from us. Since I moved here I've read her work over again, and really appreciate the light, and the humour, that I couldn't see when I was at school. I've become a fan. The way she describes growing up in these quarrying communities at the turn of the century. It's actually very modern writing. Not a word is wasted: like a miniature painting, not a brushstroke out of place.

'This is where she used to pick … er … *hazelnuts.* Had to think of the English for a minute! And blackberries. She mentions it in her autobiography, and in a disguised way it appears in some of her short stories. Her description is so precise, there's no doubt.'

Siôn and Non were brought up in the Pumlumon area in Ceredigion. They went to same secondary school, but didn't meet until, coincidentally, both had moved to work in the

* *Bicell.*

† Born Rhosgadfan 1891, died Denbigh 1985.

Bangor area. 'My granddad preached in the chapel at Ponterwyd, where I'm from. At school everyone said I was from the sticks, but up until then I thought it was the centre of the universe.'

I have the impression that here in rural Wales there is some special connection between people and nature; I ask Siôn whether this is true, or just a romantic idea on my part.

'Well, you know, last week I was asked to give a talk to Cymdeithas Hanes y Tair Llan, the local historical society. They wanted a talk about the curlew, of all things. There were people like Dafydd Wigley* in the audience. So, it was a 'history of the curlew in Wales' talk. At one level, there is a kind of historical inquiry to be made. It's a history of expansion through the first half of the twentieth century, from the mountains into the lowlands, but it's a story tucked away in local avifaunas that you need to tease out. What's better known is that it started to decline after World War Two, then there's been a cataclysmic decline from the 1980s.'

I could sense that this succinct account was merely the unravelling point for a whole Celtic knot of ornitho-historico-cultural ideas.

'Actually, you could write a history of Britain and Ireland just by tracing the local names for the curlew. The first mention of the species is in the Exeter Book, in the poem 'The Seafarer', where it has the Anglo-Saxon name *whulpan*. That name persists in north-east England as *whaup*. In Shetland and Orkney there's *spowe*, which is very similar to modern Norwegian. In Irish and Manx, and also some southern Scottish islands, it's called *crottach* – hunchback. Then in the Outer Hebrides, *guilbneach*, which is similar to the North Walian. *Curlew* is the youngest of the names, from Norse via Norman French. To this day, in Pembrokeshire, which was ruled by the Normans, the Welsh name for curlew is *corlean*.

* Dafydd Wigley, Baron Wigley of Caernarfon, leader of Plaid Cymru 1991–2000.

'I've got this great book at home, *The Linguistic Geography of Wales*.[8] My father was a librarian at the National Library of Wales and it was his favourite book. It was compiled from a … er … *holiadur*, a questionnaire sent to older people who still used local and regional words. There was a set list of items and you had to fill in which words you used for each. The list included lots of nature words but only two birds, the curlew and the magpie.

'So for the curlew, *gylfinir* is the name all around here in North Wales, and is now the standard Welsh word; it refers to the bird's long beak. But it's *chwibanogh* in Ceredigion, that's the word I grew up using, it means "whistling".'

'I remember when you would buy a book about birds and many of them would have at least two names as standard,' I say. 'It was always "lapwing or peewit", "corncrake or landrail", "meadow pipit or titlark".'

'They had a first stab at standardising Welsh bird names in the seventies. Recently there's been an increase in the use of standard names, which basically means choosing one local name over another. You grow up using one word, and of course you know the English name as well. Then you have to learn a new Welsh word if you don't come from the place where the so-called standard name is used.'

We turn right off the railway track, pausing at the scratchy song of a redstart, newly arrived from the Sahel and tentatively laying claim to a handful of oak trees, the last ones before the view opens out and we find ourselves among small, stone-walled fields. We walk a few yards uphill and look back to take in the panorama far below, dominated by Anglesey. We begin from the left, the west, and Siôn points out the landmarks, starting with the hill fort of Dinas Dinlle, a wedge-shaped hill at the shore of the Celtic Sea. Between here and there are 5 miles of small oak woods, smaller fields and the villages of Llandwrog and Llanwnda.

'Dinlle refers to Lle, one of the most famous characters in the Mabinogion. He plays an important role in the myth that is constantly used to justify the view that wren-hunting was a tradition in Celtic lands.'

'You mean like on St Stephen's Day? I'd always understood it was practised in Ireland until relatively recently.' I had in mind the former tradition of the hunting of the wren on 26 December, still known as Lá an Dreoilín (Wren Day) in parts of Ireland.

'The wren, the owl and the eagle are major characters, common to the Mabinogion and the King of the Birds story. In the Mabinogion the wren doesn't die, and isn't referred to as a king. I think wren-hunting was probably brought to England by the Normans, but soon died out. There's little hard evidence of wren-hunting in Wales or Scotland; maybe in Pembrokeshire, which was under Norman rule. William Strongbow, the Norman Earl of Pembrokeshire, invaded Ireland, so maybe it reached Ireland via Sir Benfro Saesneg[9] – English Pembrokeshire!'

We continue scanning the panorama. A degree or so east of the *dinas* is Morfa Dinlle: *morfa*, Siôn explains, means coastal land, and he reminds me that the land there is a grazed wet pasture managed for lapwings by the RSPB. Then Fort Belan, Aber Menai and Y Foryd (meaning 'bay', says Siôn); followed by Ynys Môn – Anglesey – with the dark Newborough Forest, famous for its winter roost of ravens.

'That little island off Newborough is Ynys Llanddwyn.'

'I know that name,' I say. 'Isn't that where what's-her-name, the Welsh St Valentine ...'

'Dwynwen, the patron saint of lovers. Yes, that's where she went to become a hermit, and where RSPB Cymru was born.'

My year as Director of RSPB Cymru was the ninety-ninth year of RSPB's work in Wales. The following year, starting on Dwynwen's Day, 25 January, we celebrated the centenary of the first RSPB wardens in Wales, installed on the small, flat, tidal island to protect rare roseate terns from egg collectors. It is also the place to which Dwynwen, unable to marry Maelon, retired in 460 AD, vowing never to marry but to pray God's protection for true lovers.

The west-east sweep of land, estuary and sky ends abruptly with Caernarfon Castle, like a small mountain in its own right, whose buttes and crags bite at the smooth line of the Menai Strait, teeth bared at Ynys Môn. We climb further, up a cart track, right over stony *ffrydd* – the land between the low grazing and the high moor – and right again onto the mountain. A narrow lane crosses the open ground.

'These days this is called Y Lôn Wen – The White Road, which is also the title of Kate Roberts's autobiography. It actually comes from a short story – you can probably get it in English – called *Tea in the Heather.* This hill is Moel Smytho on the map, but now aka Mynydd Grug – Heather Mountain.'

There is still heather, and gorse. It seems to me that the air is filled with skylark song, but before I say anything Siôn bemoans their lack, how few there are now compared with how it should be.

'There's no sheep hefting any more. Look,' he points a little way up the hill, 'three barren ewes. It's undergrazed up there, and intensive grazing down there.' He points out a bright green 'improved' field 300 or 400 yards away, down the slope. 'That field is only used to dry off the ewes when the lambs have gone. Where are the curlew, where are the lapwing? There should be red grouse, hen harrier.'

I remember from my year in Wales that the situation Siôn describes is not confined to this small corner of Snowdonia.

'True, but this is an SSSI* on the edge of a National Park. It's protected in theory, but we don't have sophisticated enough support mechanisms to allow farmers to manage it properly.' We continue uphill, where it seems everywhere has been named twice.

'This is Mynydd Mawr, which means Big Hill, but most people call it Mynydd Eliffant, from the shape.'

* Site of Special Scientific Interest.

As we begin to crest the hill, piles of slate waste appear over the rise, and as we walk between them, we see the 1,400-foot Moel Tryfan ahead. The quarry track takes us most of the way, and we find a short, steep path to the summit. Siôn calls this place Barclodiad y Cawr, which, he says, could be translated as 'Giant's Smock-Full', as if someone has been carrying a pile of rocks around in the bagged-up front of his smock and dumped them at the top of the hill. The exposed rocks that mark the summit are Cilgwyn conglomerate, almost fluid in appearance, like lumpy dough, a natural concrete of claystone, gravel and sea shells. On 26 June 1842, eleven years after the Geologists' Association first learnt of their existence, Charles Darwin stood here, studying the formations. For some, they proved the biblical account of Noah's flood, showing that the sea had risen even to here, allowing shells and other sea creatures to flourish. Others explained the mountain-top marine life as the result of the great upheavals of 20,000 to 30,000 years ago: glaciers dredging the sea bed and bulldozing the deposits uphill, then pressing down upon them to create young rock out of mud and pebbles.

We return down the grassy slope of Moel Tryfan, back to the track which winds between slate tips and past derelict slate-built huts, arriving at Cors y Bryniau, Alexandra Quarry. We walk to the lip of the quarry and look down into a gaping hole, part of which is filled with a cold-looking lake that reflects its slate walls in dark bottle-green. We watch through binoculars as, deep below us, two yellow machines are at work.

'There's been talk of them reopening the quarry. I think it's because of the market for slate chippings, judging by the machinery, which is just scooping stuff up and tipping it into lorries.'

Two choughs fly over us, calling, black crows of the mountains, their curved red beaks open like surgical instruments. It is a loud, percussive call like a round cobblestone hitting a granite boulder. It reminds me of a jackdaw's call, but one of stone to the jackdaw's hard steel.

'Pity they're working today,' says Siôn, 'I wanted to take you into the quarry to hear the acoustics. It sounds great with the choughs' calls pinging off the quarry face.' As the birds fly over, I can imagine what he means; the sound seems somehow incomplete without its echo. It occurs to me that this may be more than fanciful: it is a bird of quarries and cliffs, with a call that has evolved to resonate around its landscape, perhaps.

We top another shoulder of moorland, and the Nantlle ridge opens up on our right-hand side. Siôn takes an Ordnance Survey map from his rucksack, opens the middle few folds and turns it upside down to align the paper version with the ridge we see in front of us. The map is annotated with Siôn's handwriting – local names he has collected during conversations with farmers and fellow walkers, small cloughs and streams, minor peaks and cairns, whose names are unknown to the map makers.

'Do you think these names are in danger of disappearing? Do people simply not have a need for this level of detail any more?' I ask.

'Not only place names, but general terms for features in the landscape too. Take that mountain.' He points to a saddle-shaped mountain in the middle distance with what looks like a large bite taken out of it, a feature I would label with a Scottish term: a corrie.

'My grandmother had words for all the parts of a mountain, like it was an animal.' He strokes the skyline with his hand: 'There's the head, the back – all in Welsh of course – the arm – which is actually the old word for "leg" in Ceredigion – the foot, the armpit – which would be called a *cwm* if it has something built in it, like a sheep pen – and the belly.'

23 April, Shelley, West Yorkshire. 53° 35' N. 'Light thickens; and the crow makes wing to the rooky wood: good things of day begin to droop and drowse.'*

Yesterday I got home in time for the rooks' evening passage over The Hamlet and towards a line of high ground to the north, where the rookery sits in view of our house, in a clump of tall sycamores by the church. Their voices welcome and soothe; Roger Deakin likened it[10] to the roughest of folksong. The escarpment, or shelf, gave the village its Saxon names: *Scylf* in Anglian; then, by Doomsday, *Scheulay*; today, Shelley. It looks down and across to our hamlet, a mile off by rook-flight, then up to another rise over which is Denby Dale – the valley of the farm of the Danes. On the sweep of high ground that links the two is Cumberworth, the enclosure of the Celt.

Shakespeare is all over the radio this morning, the 400th anniversary of his death. He was a careful recorder of the seasons:

> daffodils,
> That come before the swallow dares, and take
> The winds of March with beauty; violets dim,
> But sweeter than the lids of Juno's eyes†

The Hamlet is six houses built of sooted sandstone, where Jane and I have awaited swallows for the past seventeen springs. We came to this place, a few miles from the north-east edge of the Peak District, in February 1999, when I was appointed RSPB Director for Northern England. I had worked for fifteen years at the RSPB's headquarters in Sandy, and we both relished a return to the hills: Jane was brought up in the Marcher lands of Shropshire and I had had a taste of the Pennine country, having spent several years in

* *Macbeth*, III. ii. Rooky, meaning 'smoky'.

† *The Winter's Tale*, IV. iii.

Lancaster. Here, Yorkshire's beauty is unselfconscious, in the gap between the county's three National Parks. Always rural, and now post-industrial, a few mills and chimneys still stand, their old smoke caught in the rough touch of the stone.

Two months after we arrived, the black grouse was declared extinct in the Peak after decades of decline, when the last remnant of a once thriving population disappeared from Swallow Moss in the Dark Peak of Staffordshire. This was not just another everyday tale of the local impoverishment of nature: it had the effect of marching the southern edge of the species' English range 120 miles north. It also symbolised the breakdown of any immunity from decline that the National Parks might have enjoyed through mere national pride alone.

Once the crown jewels of the English landscape, the region's National Parks are wildlife crime hotspots, with the Peak District among the worst. The 1949 Act* that created them sat alongside the creation of the National Health Service as the twin peaks of post-war democratisation: the one provided for the physical health of all, the other, the nation's sense of well-being through connection to nature. Five years later, in 1954, the Protection of Birds Act was passed. In its way, this also had a democratising role – by making wildlife the property of all. The great majority of people didn't trap finches for profit, or make private fetishes of rare birds' eggs. The old superstitious fear of the birds of the night, and belief in the malevolence of raptors, had died out among ordinary folk, who saw in the flight of a peregrine a thing to inspire awe. The 1954 Act codified what society already believed; only a small minority maintained the old ways and prejudices.

In 2006 we published a report, *Peak Malpractice*, which mapped the changing distribution of raptors during 1991–2005. It showed that in 1999, goshawk and peregrine nests failed across the grouse-shooting parts of the Peak, in what

* National Parks and Access to the Countryside Act 1949.

we now know was a co-ordinated assault in the war against the hated foe. Reports of suspicious activity had been on the increase: a whole raven's nest removed; an active goshawk nest destroyed, for which a gamekeeper in the Derwent Valley was eventually prosecuted; an armed man approaching a peregrine's nest; bait dosed with mevinphos, an illegal poison; peregrine chicks killed in the nest; a buzzard poisoned; another buzzard with broken legs, the stigmata of the pole trap. The report showed that at least one tradition of pre-1949 Britain had been preserved in the Peak: the systematic ridding from shooting estates of birds of prey. It took just six years of concerted effort for the goshawk and peregrine to be virtually extinguished from the grouse land.

The 1949 Act established a single Authority in each National Park, but failed to vest it with any control over the land itself. Outside the Parks, in the decade after the war, wildlife began its long decline; within the Park boundaries, there was nothing except the terrain itself to prevent the same impoverishment. Eventually that proved no barrier, and as machinery improved, no meadow was safe from the plough. It was only in the late 1990s that a determined initiative to restore the Peak's lost flower meadows was instigated. The Authority's ecologists found that 50 per cent of the very best flower-rich meadows had been lost between the mid-1980s and 1995, and overall 75 per cent of flower meadows had gone during that time, on top of the destruction wrought throughout the preceding decades. The bird most associated with upland hay meadows, the linnet's unshowy first cousin the twite, only survives in intensive care, where the RSPB's farm business advisers have targeted individual fields and helped farmers apply for special grants.

New initiatives are at last making up for lost time. Since 2003 the Peak District National Park has been working with landowners on a project called Moors for the Future. On the highest ground, the Peak District had been particularly affected by 150 years of atmospheric pollution from the industrial cities that surround it, raising acidity levels in the sphagnum bogs and causing severe peat erosion. Throughout the twentieth century the damage was compounded by drainage, mainly to increase the dry heather habitat that grouse prefer. The new project is repairing the peat bogs by blocking drainage gullies and restoring vegetation.

Elsewhere the RSPB has been pioneering integrated water catchment management with United Utilities, the water company for north-west England, who own large areas of uplands in the Pennines and the Lake District. The Sustainable Catchment Management Programme (SCaMP) takes a whole-landscape view of the uplands, and provides farmers with grants to manage livestock in ways that benefit water management and wildlife. In the Lake District, Wild Ennerdale is reimagining a landscape where natural river flow and low-density grazing will determine how the valley will evolve in the future. For the first time since 1949, a purpose for National Parks and other upland areas is starting to emerge.

In contrast, the Peak District Bird of Prey Initiative has been officially declared a failure by its chief architect, the National Park Authority's Rhodri Thomas. In the year following the publication of *Peak Malpractice* there was a further increase in wildlife crime. This included a peregrine who, X-rays showed, had survived one shooting earlier in her life, but was killed in her second encounter with a shotgun. Goshawks and peregrines 'disappeared' halfway through the breeding season. Two pairs of hen harriers nested – the first in the Peak for 138 years. Both males 'disappeared', although thanks to volunteers providing extra food, the females kept going and raised five chicks each. Over the years, Rhodri Thomas has been patiently bringing disparate interests round the table to deal with the Park's shameful lack of raptors. Of the four species most reviled in shooting circles – hen harrier, short-eared owl, goshawk and peregrine – the fate of the peregrine would be the real test of the success of the initiative. Hen harriers rarely nest at all here; short-eared owls, which are plentiful only in good vole years, are too unpredictable to draw any conclusions from their numbers. For the goshawk, the parties round the table never did agree a target. Peregrines, on the other hand, are doing well across the country away from shooting areas, and are becoming a familiar sight in inner-city Sheffield and Manchester, and in White Peak, the limestone country where there are no grouse moors. For the Dark Peak, the agreed 2015 target of fifteen pairs averaging just over two fledged chicks each seemed readily achievable. But only four pairs nested last year, only three of which produced young. This year, seven pairs have nested, but five pairs failed before fledging any young.

Two months ago, on the morning of 24 February, two birdwatchers were out walking on the Park Hall Estate moors in the heart of the Dark Peak. One of them saw a grey raptor with black wingtips perched on a distant clump of heather, and tentatively identified it as a male hen harrier. The two men set up their telescopes, and as they were focussing onto their target, one of them called out: 'An armed man dressed in camo has just jumped into the heather 20 metres from the bird.' They tried to work out what was going on: male hen harriers would never allow someone to get so close, yet the bird remained completely unperturbed. They started to film as the events unfolded, using the technique known as digiscoping – holding a phone camera to a telescope. At over half a mile's distance, the film is grainy, but the shots, coupled with the men's eye-witness account, were clear enough. After a few minutes, possibly having spotted that he was under surveillance, the man got up, walked over to what was by now clearly a decoy harrier, tucked it under his arm and left.

It was, of course, impossible to identify the armed man, and impossible to mistake his intention to lure and kill a real hen harrier, the Most Wanted of all the raptors in gamekeepers' sights. I expected an embarrassed Moorland Association to make a condemnatory statement. This is what Amanda Anderson, the Association's Chief Executive, actually said in a statement circulated by email:

> From the clip, it is very difficult to make out any detail at all, either of a person or a decoy. The identity of any person allegedly filmed is unknown, as is the location. No crime has been committed as far as we can see. Making judgements based on assumptions of the content of this clip, or indeed the intentions of those who have produced it, would be pure supposition and not something we are going to enter into.

After decades of under-performance, the National Parks are taking belated steps towards a vision that puts nature, and people's connection to it, in the foreground – for the nation. But the tripwire is always set in the same place – the private moors where the purpose of the land has never been in doubt: shooting birds for pleasure. Red grouse management is spectacularly successful at its single purpose. Across 76,000 acres of the Peak District and 64,000 acres of the North York Moors, estate managers and gamekeepers have learnt to fine-tune the habitat to make it perfect for this one species. They have argued, with some justification in the past, that other species like curlew, dunlin and golden plover are a happy by-product of their dedication and investment. That is changing fast. Recent studies show that grouse management has become so intensive, using the philosophy of intensive agriculture, that monocultures of grouse are produced at the expense of any other purpose for the land.[11] The claim that grouse management benefits other wildlife no longer holds true.[12]

Swallows have been here as long as we have, and without doubt for as long as the barns behind us have, some 350 years. In our first spring, a swallow arrived on 7 April. It was a typical appearance: a male circled low around The Hamlet, repeatedly quartering the few yards that separate the back of our house from our neighbour's barns. Nearly three weeks later, I bumped into another neighbour. He was outside, looking up at the sky.

'Twenty-sixth of April,' he said, 'swallows should be here.'

'Yes,' I said, 'the first one arrived on the seventh. Look – there's one.'

'Well that's odd, they are supposed to arrive on the twenty-sixth.'

Since then, swallows have arrived here on dates ranging from 3 to 15 April. In every year, from the day the first bird has arrived, there have been swallows every day until the end of the summer. This year has been the only exception. On 4 April, I saw two from an upstairs window. The following day and each day until I returned to France on the 7th, I looked for them in vain. Each evening on the phone Jane would report news from The Hamlet, and the continued absence of swallows. They may have arrived after I crossed the Channel into England and Jane came south to meet me, when neither of us was here to witness it. We arrived home on the 15th, late in the evening, and saw a pair the next day.

I have noticed that the first bird back is always a male, or a male with someone I presume to be his mate; never a lone female. Unpaired males arrive back on their breeding grounds first, singing as they come, to find suitable habitat, familiarise themselves with the lie of the land and wait for an unattached female to turn up. Established pairs arrive together. Ringing studies have shown that as long as both partners survive, they remain faithful to each other (it would be better to say they remain together, despite regular infidelity) for life, and generally faithful to their home ground. Unknown until recently was whether they stay together during their long migrations and on their wintering grounds. In July 2012 a pair were fitted with geolocator tags* at their breeding site in northern Spain. Both left the area on the same day, 9 September, and arrived in the same wintering area in West Africa on 20 September. The return northward journey also started on the same day, 20 March, and both individuals were back at their Spanish breeding site on 10 April. Throughout the winter, wherever they went, they remained together.[13]

25 April, Loch Kinord, Aberdeenshire. 57° 05' N. The first willow warbler arrived here on 19 April, the noticeboard in the visitor centre says, and there are four or five singing in the woods at the edge of the loch. Its simple, down-the-scale whistle is the confirmation I listen for that spring has arrived, and I am pleased to hear a little community of them at last.

Since 1985 they have declined by 8 per cent if measured at UK level, but in England the loss is 40 per cent – and twice that much in east and south-east England; conversely, it has increased by 23 per cent in Scotland and 73 per cent in Northern Ireland.[14] One possible explanation is that climatic warming and wetting is allowing this exclusively insectivorous species to inhabit more upland areas, but not fully offsetting declines due to dryer conditions in south and east England. Another possibility, suggested by analysing the chemical make-up of feathers grown in Africa, is that the different populations winter in different places, and the southern breeding birds experience more severe pressures while away from their breeding grounds.[15,16]

* A tiny tag that periodically records ambient light to determine location. The absence of any telecommunication device enables it to be fitted to very small birds; it must therefore be recovered before the data it contains can be downloaded.

Like all long-distance migrants, the willow warbler is here because it gradually adapted its life cycle to the new opportunities presented by the ending of the last ice age. 11,500 years ago, Scotland's glaciers were in retreat. The ice cliff at the edge of one that ground its slow journey over this granite started to collapse. The foot of the cliff was melting faster than the rest, creating a deep cave that threatened to undermine the great tongue of ice above it. Clefts appeared across the glacier, as its unsupported front portion started to lean forward, until with a roar that shook the earth, the glacier calved, and the ice calf itself cracked into two bergs, each a mile wide. They became mountains of ice in an accreting landscape, as sediments formed around them. Their own weight pushed them deep into their surroundings, until slowly they, too, melted. The holes they left behind filled with sediment and water, and were given names by the Mesolithic people who came after the ice. Whatever those names were then, today they are Loch Davan and Loch Kinord, known to geologists as kettle holes.

It has been sunny, but it starts snowing as we enter the birch wood at the edge of the loch. On our left, the wood opens out and gives way to a few acres of rushy pasture. Its name is Bogingore, the little bog of the crane. Some of the birch trees have lost their crowns, snapped clean off in past storms. In this snow-light, this strange pall of a million flickering pixels seen through the snowfall's lacework, these decapitated trees have a regal presence. One in particular, in the centre of a slight hollow, appears out of the veiling snow like some ageless stele, unbending to the pressure of wind, bedecked in regalia of lichen and polypore.

We stopped here, at the edge of the Cairngorms National Park, mainly to find goldeneye, a beautiful duck of the taiga which has nested in Scotland since 1970. We have come to see their spring display, which shows off the drake's black and white plumage to stunning effect. If the snow doesn't ease and the sun doesn't reappear soon, the chances of finding them in a mood for courtship are slim, and I will have to wait until I reach Scandinavia to get another chance. I look across the lake for any sign that the sky might be clearing, and amid the murk a figure flies high over the water, a vague grey silhouette – but recognisably, thanks to its gull-like flight on broad, angled wings, an osprey. It is sign enough that we've seen the best that Loch Kinord can offer today, and we turn back. We have the whole of the Cairngorms National Park ahead of us, and if the snow continues, we will have a hairy drive through the eastern Grampian Mountains.

On the way back to the car we flush a woodcock from the ground close to the path. It flies low and fast and jinks between trees, its wing quills whistling. I study the spot of ground it rose from: pale-brown bracken fronds pressed into dark-brown peat, dull orange moss, spent grass stalks: all the colours of a woodcock, intricately patterned like the bird. Before we leave the woods, the snow stops abruptly and a minute later the sun is out. In sun, we notice patches of primroses we hadn't seen before, and hear chaffinches and chiffchaffs again, as if spring itself were welling up, irresistible to the too-feeble pressure of the retreating winter.

I have heard that there is a gowk stane near here, and we detour north along the eastern edge of the Muir of Dinnet to find it. Until a few days ago, I had never heard of a gowk stane. After I woke under that bright fenland sky, I travelled the few miles to Cambridge for a one-day conference on the science and lore of cuckoos. There I learnt that in Celtic Britain and Ireland, certain prominent rocks, usually the so-called erratic boulders transported from some distant origin in the tumult of a glacier, are given this name. *Gowk* is a Scots word, derived from the Old Norse *gaukr*, or the Anglo-Saxon *gouk;* a word which in England was replaced by the French *coucou* when the Normans arrived. The cuckoo is a favourite subject of Celtic mythology, and one explanation of the gowk stanes is that they are stones from where the cuckoo calls first on arriving in spring.

We find a boulder about 6 feet high, roughly the shape of a rounded cube. It is sitting at the edge of a hay field, where the birch trees have encroached to form a young copse a few yards from a track. There is no sign to tell us that we have found the gowk stane, but I imagine it standing proud in its unwooded surroundings in the first millennia of the post-ice age. It will have been accorded status by all the peoples who came after the glacier whose remnants formed Lochs Kinord and Davan.

At the Cambridge meeting, Dr Chris Hewson of the British Trust for Ornithology announced the extraordinary findings of his latest research.[17] Satellite tagging had enabled him to track forty-two individual cuckoos from nine different breeding areas in Britain to their wintering sites in Africa. He found that they all migrated to the same area in Central Africa, but the route they took to get there differed. Some took a shorter route via Spain, some a longer one through Italy or the Balkans. Hewson compared the proportion of birds from each of the nine populations that took each route, and found that many but not all of the birds from England took the western route through Spain, while Scottish and Welsh birds tended to head east through Italy. In England, the proportion of east-route and west-route birds in each population varied. He then compared known population trends between the nine areas, and found that despite the western route being shorter, the proportion of birds using it strongly correlated with population decline. I was surprised when Chris explained that most of the increased mortality occurred before the birds even reached Africa. West-flying birds seem to leave later, spending more time fattening up in Britain and stopping to refuel fewer times during their shorter migration. The decline in large moths, whose caterpillars are the main food of adult cuckoos, is more pronounced in the places where west-flying birds form a higher proportion. This, along with the increased frequency of drought in

Spain, are suspected causes of the rapid disappearance of the cuckoo.

In the next decade or two, the loss of the cuckoo from much of Britain will be the defining cultural bereavement of the age. The last such extinction was the wolf, which, for almost everyone, was already known only in folk memory when it was finally extirpated. Officially, this was in 1680, not many miles away. According to local legend, though, the last wolf was killed even nearer to here, by MacQueen of Findhorn in 1743. In any case it was, over most of the country, only a haunter of fables, superstitions, fairy tales, poems, epics and allegories. For centuries before and after 1680, very few people would have known if it existed or not; the transition from feared and hated neighbour to myth-dog of a dark yore was long and slow. The winter months were Wolf Time, when the risk of an encounter round the sheep pens and farmsteads increased, and Norse-origin tales of wolves devouring the sun were retold by flickering flame-light.

In Wessex, when the fearful nights drew back in spring, people sang 'Sumer is icumen in'. It may not be the oldest song in the English language, but it is the one that has survived longest. It was written down in about 1260 in Wessex dialect, and its exhortation to 'sing cuccu!' must have rung out for decades, if not centuries, before that.

Loch Garten, Highland. 57° 15' N. We arrive at 4.30 p.m. and walk through the pine forest to the loch shore, where we witness snow falling three ways.

First, pea-sized hail hits the loch with a sound like the water is simmering, bubbling from below. Evenly spaced and equal-sized discs of ripple dot the surface. From the centre of each disc a water spike – a Rayleigh jet – shoots 3 or 4 inches upwards, carrying at its tip the rebounding hail, whose ice-blueness is like bioluminescence, a colony of dancing loch sprites.

After the hail, a dry, gritty snow lands without sound or ripple, and each grain continues on a refracted trajectory into the dark water, like neutrinos. Their lights fade rapidly as they melt.

Then a hole in the clouds lets an oblique sun back-light the snow, which has formed into flakes that look bright and big. The low sun projects cameos of the surrounding trees onto the loch, and in another light-dance the snowflakes' reflections rise to meet their falling partners, to kiss and die at the surface.

At the RSPB osprey centre, Visitor Officer Rob Ballinger sets up the telescope so that we can see the most famous bird's nest in the world. In 1923, the Anglo-Argentine author William Henry Hudson wrote:

> For three-quarters of a century the story of the Osprey in Britain was a story of incessant persecution. To the game-keeper it is one of the hated tribe of Hawks; to the water-keeper it is an object of still more deadly hate, because it takes of the fish in lake or stream: for the osprey feeds exclusively on fish, plunging down on them like a gannet and fixing its curved talons in them with the strike of a falcon. So certain is its aim than in the old days it was believed to possess some magical power by which fish were compelled to rise and float to the surface in readiness to be taken.[18]

It is a magical power that Shakespeare alludes to:

> I think he'll be to Rome as is the asprey to the fish, who takes it by sov'reignty of nature.*

* *Coriolanus*, IV. vii.

The osprey became extinct in England in 1840, and the last Scottish ospreys known to Hudson nested in 1904, although it appears to have attempted to breed sporadically until 1916, and perhaps into the 1920s and 1930s. Then in 1954 a pair of ospreys paused here at Loch Garten on route to Scandinavia. They found it to their liking, and became the first ospreys to nest in Britain for at least a quarter of a century. The recovery was slow at first. Organochlorine pesticides were still ravaging the food chain, and egg collectors were a constant threat. By 1976 the population was still only fourteen pairs, and this increase came only at the cost of a monumental effort to maintain a 24-hour watch on each nest. The early efforts of the RSPB marked a radical shift in the approach to rare breeding birds: in 1959 secrecy was lifted, and the public was invited in to see the birds. In the first year, 14,000 visitors came, since when each year's arrival of ospreys at Loch Garten is reported in the local and national press. Increasingly, CCTV and remote alarm systems were used, and by 1991 there were seventy-one pairs breeding in Scotland, and more than twice as many ten years later. Economists have assessed the value of the osprey to the economy of Speyside: in one year – 2004 – visitors to this one site brought £1.89 million of additional spending into local businesses.

Concern for the osprey is not new. As long ago as 1883 Henry Seebohm noted:

> Years ago, before the railway had joined the Highland solitudes with southern industry, before such attention was given to the preservation of game and the destruction of 'vermin', the Osprey dwelt among the mountain Lochs, or on the brown heathlands studded thickly with stunted fir and birch trees. Now his haunts, which are only few and far between, appear to be the dense pine-forests that clothe the steep and rocky hillsides. … Here, on these strictly preserved estates, the Osprey is a regular visitor in

> the summer months, and bids fair, with the aid of the protection now afforded it, to re-instate itself in the home of its ancestors.[19]

Seebohm may have been referring to the work of two Highland landowners, John Peter Grant, Laird of Rothiemurchus, and Donald Cameron, Laird of Lochiel, who in 1893 would be awarded the silver medal of the Zoological Society of London for their efforts to prevent the loss of the remaining few pairs, hoping to create a nucleus for the future repopulation of Scotland. Grant's land included, as it does today, the ruined thirteenth-century castle on the island in Loch an Eilein, a nest-site that had been continually occupied for at least sixty years at the time of the award.

There is a female on the nest, but we cannot see her through the scope as she is lying flat over her eggs. We can, though, see her live on a TV screen, hard snow bouncing off her back. Her name is EJ. Her mate, Odin, is perched a few yards away on the dead spur of an old pine, and we refocus the scope onto him. Even from 200 yards away his talons, designed for gripping and hauling fish, menace. The colour of steel, they close over the branch like workshop tools. The snow pelts him ineffectually; his head tilts so that he can scan the sky for rivals, and turns into the wind with no change of expression. The fringe of off-white feathers that stick out at the back of his head are like shards of ice.

26 April 2016, Cairn Gorm, Highland. 57° 08' N. There is a strange flickering against the sky; it reminds me of yesterday's white lights dancing across the dark tympanum of Loch Garten – another snow-dance. And how strange – like paper butterflies hanging in a lust-crazed breeze, fluttering on the taut wires of a mobile. Against the dark, snow-bound clouds, it is like an odd celestial light source. It lasts a few seconds.

Jane asks: 'Did you see that?'

We are halfway up Cairn Gorm, in the car park at the base of the funicular railway. There is a low perimeter wall, beyond which the mountain slopes down into Allt Coire an t-Sneachda, one of the valleys that funnel snow-melt into Loch Morlich. A dozen snow buntings appeared over the wall having flown up from below, hugging the line of the slope to ease their flight against the wind. Their sudden encounter with its full force flustered them upwards. Their wingbeats had the flicker of old film: the inner half of a snow bunting's wing is gleaming white, the rest of its upperparts dark. The gloaming and the gleam alternating frame by frame, was what we saw.

We are, in fact, halfway down, having reached the top station already, and been half-blinded by the snow. There is no visible sun today, only a thick sky dispensing miserly rations of used sunlight. But the lying snow wants none of it, and throws it back in concentrated form to sting the eyes like salt. The limit of visibility is 2 or 3 yards up there, and the temperature -16 Celsius. On the way down we were the only passengers, and we told the young funicular driver of our thwarted hope of seeing a ptarmigan, the snow grouse that turns white in winter and should be moulting back the colours of lichen-spattered granite now.

'We see them a lot of the time,' he told us, 'but in this weather they'll be down in the gullies and behind snowdrifts. Or along fence-lines. There are a few places on the way down where you stand a chance – I'll keep my eyes open.'

He slowed the train down several times, and kept his cab window open. Twice we stopped completely, and he leaned out, staring at the hillside with slitted eyes behind his ski shades; we scanned with binoculars. We saw a few red grouse lower down towards the base station, but that was all.

Last month, the RSPB made public the findings of an investigation into a male hen harrier that was found dead here in the Cairngorms. 'Lad' was fitted with a satellite tag by licensed RSPB staff on 16 July 2015, a few days before he fledged from a nest on the Gaick estate owned by Danish businessman and conservationist Anders Povlsen. After fledging in late July, Lad stayed close to the nesting area until the last week of August, when he then moved a short distance away from the estate. A few days later, on 3 September, RSPB staff monitoring the transmissions from Lad's tag became concerned that he had stopped moving; he stayed in one spot on a grouse moor near Newtonmore for a week, obviously dead. RSPB staff went in and found the body, which they took to Scotland's Rural College's veterinary laboratory near Penicuik the following day. The preliminary post-mortem report stated:

> The skin was split open on the left side of the neck parallel with the jugular groove. There was haemorrhage in the sub-cutaneous tissues in this area and a horizontal split in the trachea. There was damage to three feathers of the right wing consisting of a single groove mark perpendicular to the shaft of each feather. The body was X-rayed, and the report came back: Despite the failure to identify metallic fragments within the carcass the appearance of the damage to the wing feathers is consistent with damage caused by shooting. The injury to the neck could be explained by a shot gun pellet passing straight through the soft tissue of the neck. Both injuries could have brought the bird down and proved fatal.

The Cairngorms National Park was established in 2003, since when there have been over sixty recorded incidents of illegal raptor persecution. In the last five years – between 2010 and the death of Lad – the catalogue of victims and methods reads like a Victorian Gothic chamber of horrors:

six buzzards poisoned and one rabbit bait – laced with the illegal poison Aldicarb – found; one buzzard, one golden eagle, two short-eared owls, one peregrine and two hen harriers shot; one goshawk killed in a spring trap and two more traps found; one peregrine's nest burnt out, a buzzard's and a goshawk's nest shot out and one white-tailed eagle nest tree felled. And four satellite-tagged eagles (three golden and one white-tailed) *disappeared*, in the Latin American sense.[20]

The war on wildlife is not restricted to birds of prey. According to leading upland ecologist Dr Adam Watson in his 2013 book *Mammals in North-East Highlands,* grouse estates in the Cairngorm National Park have been targeting another enemy species: the beautiful and innocent mountain hare. Its crime is to be a host to ticks which may then attach to grouse, passing on a fatal virus known as 'louping ill'. Last month, photographs showing low-loader vehicles piled high with hundreds of hare corpses killed in the Cairngorms confirmed Watson's view that the species is being systematically persecuted.

Over 40 per cent of the National Park is covered by grouse moors, including at the Royal Family's Balmoral estate, which is not untainted by upland shoots' cavalier attitude to the wildlife in their care. On 17 June 2009 a gamekeeper on the estate was fined £450 at Stonehaven Sheriff Court after a badger was killed by one of his snares. The badger had lain there for at least two weeks, when the law requires that snares, which are set for foxes, be checked daily. It is illegal to set traps to kill badgers, and there was no proof that this was the keeper's intention. His snares were placed near a 'stink pit', where animal carcasses are dumped into a small pit to lure foxes, which are then snared legally, if indefensibly.

Sweden

27 April, Varberg, Halland. 57° 06' N. We left Abernethy early, in case our drive back to Edinburgh was slowed by ice. The Monadhliaths caught the early sun and gleamed over the first 28 miles of our journey south. They are the westernmost range of the Cairngorms National Park, and stand like barbicans at the gates of the Highlands. To drive alongside them, along the A9 – once a military road built to prevent a repeat of the Jacobite rebellion – is to be escorted from the granite citadel by four Munros, silent, patient but eager to close the gates behind you. They are ice-white in the morning sun; under the snow, even now, the budding heather is weaving for them a purple summer livery. But their name means 'grey mountains' in Gaelic, and behind their staunch ramparts lie murky lands, where adventuring eagles meet their doom.

In 2010 a golden eagle and a white-tailed eagle were found dead in the Monadhliaths. Post-mortem tests showed they had been poisoned with carbofuran – a substance for which there is no legal purpose, and the favourite poison of those who seek to rid the moors of birds of prey. Since then, RSPB Scotland, Forestry Commission Scotland and the Highland Foundation for Wildlife have tracked eagles with satellite transmitters fitted prior to their first flights. In November 2011, a golden eagle that had been tracked for sixteen months disappeared suddenly in the Monadhliaths. It was last recorded in the hills above Strathdearn, in the grouse moors close to where the two eagles had been poisoned the year before. Seven months later, another young eagle vanished in the same mountains, and then three more the following year, 2014.

At midday I dropped the hire car off at Edinburgh Waverley, waved Jane off on her train to Yorkshire, climbed onto the upper deck of the airport bus and sat at the front. As the Georgian architecture of the city gave way to prosaic suburbs, I checked my emails. There was one from Joan, my friend at Wicken Fen. The subject line read simply: 'Ralph'. I hesitated for a while before swiping the message open. Ralph Sargeant had been unwell for some months when I met him a few days ago, and was due in hospital for more tests. Joan's message was brief: 'Unfortunately I have to tell you that Ralph died on 25th April.'

From Edinburgh, I flew to Copenhagen, and caught a direct train from the Danish capital to this small port on the Kattegat coast of Sweden. A slow dusk fell during the journey, and at each stop the train disgorged another few passengers. After Halmstad the carriage emptied apart from one other person. I retrieved my rucksack from the overhead rack and took out my notebook. I spent the last half-hour of the journey reading the notes I took during my conversation with Ralph, and during the walk around Wicken Fen and the hours spent listening to the Ouse Washes at night. Now, in a small guest house on the main street in Varberg, I am thinking of Ralph, and that generation of fen-men whose life on the land straddled the last wave of intensification that finished off a long transformation begun 300 years earlier. My father left the Fens for national service, and never returned to live there. His pre-war memories, and Ralph's later ones, would testify first-hand to two of the changes that define modern rural Britain: the impoverishment of landscape, and the concentration of wildlife within small pockets of land set aside for the purpose.

28 April, Getterön, Halland. 57° 07' N. I watch them arrive; I am standing on a rocky knoll, they are already gliding, making their final approach, so I cannot tell from which direction they have come. They are too far downwind for me to hear their call but I know they will be calling to each other – they always do. They drop into the rushy meadow between the road and the water's edge 400 yards away and will be calling to each other because they have settled uneasily. A few are unable to resist their instinct to dance, and leap with the springy force of their powerful leg muscles, then flap for a few yards with a false amateurishness like practising fledglings. They corrupt the stately poise of the main flock; it is not like in La Serena, where winter cranes fed, moved and flew in serene groups. Even at this distance, I sense tension. Eventually they settle to feed, and a swallow darts into the foreground and flies towards me, suddenly banking left – my right – and vanishing.

This is Getterön, a nature reserve on the west coast, the Kattegat Sea, which was created accidentally in 1936. Getterön was originally the name of an island, and the island is now a peninsula, formed by infilling the stretch of water between it and the mainland. This created two bays out of one: to the south, the port of Varberg, where I am staying, could now be expanded; to the north, new coastal drift and sedimentation patterns along with dredgings dumped here from the port created 850 acres of salt-marsh, grazing marsh and reed bed.

Having watched the cranes arrive, I move to a platform overlooking the coastal marshes, an elevated vantage at the edge of the rocky former coastline. A pair of marsh harriers have dropped into the reeds to the left, close to the road; black-headed gulls are spaced evenly across an island in nesting array; avocets have arrived and have started to pair off; and a lesser whitethroat, perhaps on its first morning home, is chattering his territorial claim in the nearby bushes.

The commonest bird is one that is less than halfway through its journey to the north lands. Five thousand barnacle geese are distributed across the short-grass salt-marsh; here, I suppose, for a few days only. They have three main breeding areas: in north-east Greenland, in Svalbard and in the islands of Novaya Zemlya and Vaygach in the Russian high Arctic. Svalbard and Greenland geese winter in separate parts of Scotland, stopping during migration in Norway. The Russian population winters mainly in the Netherlands, flying via Germany, Sweden and Estonia. These birds, then, have another 1,600 miles to fly to their breeding grounds on the tundra. Wherever it is found, the barnacle goose has the habit of appearing from over the sea. Their English name reflects Gerald of Wales's observation that the barnacle goose and the goose barnacle have in common a long black neck (in the goose) or stem (in the barnacle) and a white body (in the barnacle) or face (in the goose). He concluded that they were one and the same, embryo and adult.

In the visitor centre, I stand with a mug of coffee in the bay of a glass wall that overlooks the marshes. I realise that I am blocking someone's view and apologise.

'That's OK,' he says, 'I'm not looking at anything in particular. What I mean is, I'm not an ornithologist like you are. I mean, you're probably an expert; I just think birds are wonderful. Are you English?'

'Yes.'

'Good, I prefer speaking English to Swedish.'

'Where are you from?'

'I'm Swedish, but I think of English as my first language. It's not really, but I prefer it. Peter, by the way, Peter Ekström.'

'Laurence Rose. Are you from around here?'

'No, I'm on holiday. I come here a lot though, to see the birds.'

'It sounds to me like you *are* an ornithologist.'

'No, I just watch them, I look at them intently. I think it's amazing to see the way they move and interact. I watch everything they do. And I listen to them even more intently.'

'Me too,' I say. 'I'm particularly attentive to the sounds of nature, as a composer.'

'Like Olivier Messiaen. I think the blackbird is the greatest composer.'

'That's what Messiaen said, too.'

'Really? I believe it. British Radio 3 is the best channel. I listen to it all the time, except I sometimes watch the debates in the House of Lords on the BBC Parliament channel. Are you a full-time composer?'

'No, an amateur. Actually, I work in wildlife conservation.'

'The RSPB? I'm a member.'

I ask Peter about his work, and he tells me that he is an inventor. He is a slender, middle-aged man dressed in cycling gear, and I ask about his trip, and why he came to be in Getterön.

'My mother is a very dominant person', he answers. It is, of course, an unexpected answer, but I am beginning to realise that his thought processes are lateral by default, which I suppose is a prerequisite for an inventor.

'She is, you know, a Jewish mother. Except she's not Jewish. I should have gone home a couple of days ago, but I'm enjoying the freedom of the road too much.' He realises he hasn't really answered my question, and tries a different approach.

'I'm trying out a new bike, a Swiss model. It's good, but I am working on improving it. I can't say much more about it, of course, as an inventor. Anyway, I love this coast and I come to this reserve every year. I taught her to SMS – how do you say SMS in English?'

'Text?'

'Yes, she text messages me all the time asking where I am. You understand, I am sixty-one. Many people have relationships like this. I never really got out of it.'

I mention the cranes I saw this morning and the role they seem to be playing in my journeys. Peter explains that every year their arrival at the famous migration stopover site at Hornborgasjön is the top news story on Swedish television.

'That was about three weeks ago, they come earlier every year.' We exchange email addresses, and when I correct his spelling of my name, he says:

'Like Laurence Olivier. I love him in *Henry V*. But Gielgud – his "To be or not to be" got me *here*.' He clutches his heart.

I walk along the peninsula road to the south-west corner of the reserve, which abuts an airfield built on the infill that created the isthmus between here and Getterön island. There's a strengthened headwind that swirls meadow pipits and linnets off their chosen flight paths. The flat sky has darkened. The mere's surface has roughened and no longer reflects anything at all; it has instead taken on a darker version of the sky's obscure tints. To my left, at the airfield, I notice a movement, a large animal of some kind 200 yards or so away. I focus my binoculars on the spot, but see nothing at first; then it reappears at the edge of the runway, not such a large beast after all, but two roe deer bucks in such close pursuit that chaser and chased can't be separated. They zigzag round the red and white airstrip markers and through a fence, threading themselves between the wires without slowing down or breaking stride. They continue across the salt-marsh, between high tussocks and plashing across shallow pans, outpacing their hoof-splash by a full two lengths. At every turn, the pursuing animal matches the other's path exactly. The chase ends only when the front animal leaps into the mere, and even here, the chasing buck continues for 20 yards or so, the deepening water slowing them both down to half the speed, until the victorious one is satisfied the usurper is banished from his territory. After twenty seconds, with the coast apparently clear, the chastened one begins to walk back to the edge of the mere, but changes his mind, and instead heads east across the water, at times swimming across deep stretches with only his head visible, at other times walking in water only fetlock-deep. After about ten minutes he completes the 450-yard crossing and disappears into a fringe of reeds close to the visitor centre.

I continue west past the airfield, and walk the two and a half miles to the tip of the peninsula. There are smoothed rocks the shape of slumbering seals, all with the same pattern of north-south aligned glacier-scraped scars. In the shelter of the rocks, junipers and *Rosa rugosa* in turn give shelter to feeding willow warblers and goldcrests. At the convex top of the old island there are mini-bogs of sphagnum and heather. By the time I reach Gubbanäsen, the outermost headland, the wind has acquired a cargo of rain. The irregular coastline has put a turmoil into the wind, and none of the rocks has a lee against it, so I pull on waterproof trousers, pull the hood of my waterproof jacket as far forward as I can, and wait it out. One boulder the size of a small house leans back at an angle I can recline against, and has a companion rock that reclines at the opposite angle, perfect for a footrest. I watch the sea, which is calmer than the air above it and is the same deep slate colour as the sky, and listen to the rain bouncing off my clothes. Lights form on the sea – drake eider ducks whose ivory plumage reflects a sudden spillage of sunlight that has burst through a split in the clouds. With no let-up in the rain, and the source of the light hidden, it is like watching a flotilla of bobbing bumboats switching their lights on one by one. In total a hundred are revealed, almost all males, drifting towards the shore. When the clattering rain suddenly stops, I hear their gossip-mongering 'oo-ers' instead.

The walk back to Varberg is under a different kind of sky, with a yin and yang design formed by the dark shower belt that has now passed over and a portion stretching to the south and east that promises a brighter spring evening. Most of the barnacle geese have congregated at the airfield, grazing in densely packed assemblies. Each bird takes its turn to lift its white face and join the 5 per cent or so who form a collective vigil at any one time. The effect is a kind of slow shimmer, like the massed blinking of reflections on the sea, an image enhanced by the arrival of another 500 geese folding into the rest like a slow breaker wave.

I return to the rocky knoll hoping that the cranes haven't moved on. The white yang-clouds have dispersed and the sky is mostly clear, and I reach the top of the hill in a warm-coloured early evening light. The cranes are still in their field, apart from six, who have taken over the adjacent field as a dancing arena. They are close together, but seem to be dancing to themselves, paying no attention to one another. One bows low to pluck some grass, its wings flapping loosely; one flaps its wings with just enough down-thrust to make its jump a slow-motion one, springing up each time it lands, and completing a circle in four twisting jumps; one jumps and flaps in order to stick its legs out horizontally; two flap their wings without becoming airborne at all, simply running for 10 or 20 yards. Then suddenly they stop, and feed together peacefully for a few minutes. Twenty birds from the main flock take to the air and drop in to join them. Two of the arrivals land apart from the rest and dance to one another. Theirs is a stately dance: one, the male, stands with his head raised and his wings in an angel pose, while his mate lifts her head and folds her neck back over her wings, so that her head nestles in the ruffed bustle formed by her extended inner wing feathers. I cannot distinguish their calls from among the wind-muffled chorus of the flock, but their posture is that of the 'unison calling' display, involving a complex series of co-ordinated calls.

And then they take off, the whole flock, group by group, in a seemingly disordered slow swirl at first; but gradually, according to some unfathomable system, within half a minute they have fallen into a formation and head east over the fields, inland.

29 April, Höryda Gård, Västra Götaland. 58° 18' N. A glimpse of a red fox sitting in a field as I passed by on the train made me think of the vixen that lives at The Hamlet. The spring of her year began in early March, for by the night of the twelfth her motherhood was plain: eight gentle swellings parted the fur of her belly as she foraged in the field next to our house. By this time last year she had at least one tiny cub, when I saw her on 10 April, carrying it by the scruff of its neck. I suspected she was moving dens, transporting her offspring one by one. For a few seconds, one of her trips brought her within range of a night-time camera trap I had set up at the edge of the field.

There, just outside Falköping, in plain daylight, was such a *red* red fox: *Vulpes vulpes vulpes,* the Scandinavian fox. It was bigger and redder than the *Vulpes vulpes crucigera* foxes that visit our fields, which are in turn bigger and redder than the *silacea* Iberian fox I watched in Belchite a few days after I last saw our vixen. However much these spring journeys may have bonded me to the crane, the fox is my permanent totem.

For one so ubiquitous in folk stories, myths and reality, it is a difficult animal to know. The Hamlet foxes appear to me mainly as monochrome figures that emerge from darkness into pools of infrared light to feed. By watching foxes hunting unseen prey, in 2011 Czech scientists discovered that while hearing played a crucial role, success depended mainly on the animals aligning themselves with the earth's magnetism.[1] Foxes are somehow able to 'see' the magnetic field and use it for range-finding before pouncing on their prey. Such adaptation puts its powers beyond our ken; what is sensory for a fox is extra-sensory to us; not supernatural, but super-natural.

The fox is the most enduring character in European folk tales. Unlike the wolf and the bear, which instilled fear and acquired enemy status, foxes have always been treated as cohabiters, sometimes unwelcome, but animals we have had to come to terms with as they move among us. Many tales make lovable rogues of them; even those that characterise them as sneak-thieves and cheats stop short of outright vilification and profess a degree of respect. No folk depiction, ancient or modern, contributes much to our understanding of this uncanny beast. Fox as trickster says nothing beyond the simple ecological truth of a highly adaptable omnivore. The rest is made up, be it the unsupportable libel of malevolence or the understandable illusion of magical power.

My two favourite pieces of fox literature are based on direct knowledge and experience of the animal, but make no pretence of understanding. Instead, mystery and miscomprehension are the starting point for inner reflection, for a teasing out of the authors' own self-awareness.

Rudolf Těsnohlídek's *The Adventures of Vixen Sharp-Ears* appeared as a serialised comic strip in the Czech newspaper *Lidové noviny* (People's news) in 1920, inspiring the seventy-year-old Leoš Janáček to write one of the twentieth century's best-loved operas. Before writing his own libretto and embarking on the music, Janáček studied the animals of the forest with Těsnohlídek's help. The Czech word *bystroušky*, sharp-ears, has a double meaning, synonymous with cunning. *The Cunning Little Vixen*, as the opera eventually became known in English, transformed the originally comedic cartoon into a philosophical reflection on the cycle of life and death and desire for a return to simplicity. A curious vixen cub chases a frog, which lands in the lap of a sleeping forester, who captures Sharp-Ears and takes her home as a pet. As an adult, Sharp-Ears escapes, but is killed by a poacher who gifts her pelt to his fiancée, a gypsy girl with whom the forester is infatuated. At the poacher's wedding, the forester discovers that his lost pet fox is now dead, and sees the object of his unrequited love married. He returns to the place where he first met Sharp-Ears and sits at the tree grieving his double loss. As his grief grows, a frog unexpectedly jumps into his lap, the grandson of the one who appeared at the start of the opera. This reassurance of the cycle of death leading to new life gives the forester a sense of deep peace.

At Janáček's request, the final scene from *The Cunning Little Vixen* was performed at his funeral in 1928. The old man had spent years infatuated with the married and much younger Kamila Stösslová, and in the opera's simple folk tunes and allegory (a departure from the dark complexity of his usual style), he makes his deeply personal peace with the world.

Ted Hughes's 'The Thought-Fox' is a poem about writing a poem. In a room late at night the poet is sitting alone at his desk. Outside the night is dark and silent, but the poet senses a presence 'entering the loneliness'. The night is the darkness of the poet's imagination, out of which a vague idea emerges. It has no clear outline; it is not seen but sensed; it is compared to a fox, delicately sensing its way through the undergrowth. The fox emerges only slowly out of the formlessness of the night. The fox is the poem, and the poem is the fox: 'The page is printed.'

The bus from Falköping drops me at Härlunda, a hamlet on the road to Skara. From there, there is a mile to walk along a wych elm avenue to Höryda Gård, the farm where I will be staying for two nights. The trees are a-chatter with fieldfares and redwings, and halfway along the avenue I meet Ewa, the farmer, who takes me to my ground-floor room. The view looks back down the line of trees and across a field of beef cattle, where Ewa is back, her blonde pony-tail dancing as she pushes Hereford bulls around. Fifteen feet from my window a fieldfare is finishing her nest, daubing its straw floor with mud. She brings it beak-load by beak-load, smoothing each contribution with her breast, spinning and waggling like a honey bee.

30 April, Hornborgasjön, Västra Götaland. 58° 19' N. The fieldfare is still building her nest, and Ewa is making quiet kitchen sounds. She tells me that there is a track into the forest beyond the fields to the west of the farmhouse and I am welcome to go there.

It is the kind of wet forest I imagine once covered much of Britain, of aspen, birch and spruce. There is a narrow ditch at the trackside with the odd clump of marsh marigolds clinging to it, and where the wind has thrown a spruce to the ground a black pool lies in its vacated root hollow. There are wild blackcurrant bushes growing in these wet areas, and wild raspberries on any ridge or mound standing proud of the water table. There are bog birds here, as well as forest birds. I hear a snipe drumming, and another birdcall from above and behind me; a wader of some kind crosses the gap in the trees above the track as I look up. It calls again. It is a clear call, with pitch and rhythm that could be written down; I think it is a wood sandpiper, and I mark the symbols in my book as a reminder of the repeated three notes, the first two slurred together, the middle note higher in pitch. It is often found where the fieldfare breeds, taking to the trees to reuse the thrush's nest for its own eggs.

It is a mile from the farmhouse to the bus stop at Härlunda, and the forest track leads in the opposite direction. I must return to the farm and make my way along the avenue of elms to the road. I need to catch the 11.09 to Trandansen, which is not a town but the name they have given to the bus stop at Hornborgasjön, Lake Hornborga. It means 'crane dancing'. On my way to the bus, I see two cranes in a field that is edged by a dark, dense plantation.

They call in unison, and the taut trunks echo an antiphon, the voice of the voice of the bird.

Baswenaazhi. It is the name given by the Ojibwe Native Americans to the sandhill crane, and to one of five clans into which the people are traditionally divided. It means 'echo-maker', and the echo is as much a totem as is its maker.

At Trandansen a visitor centre stands on a low hill overlooking Hornborgasjön. The lake was formed when a fjord lost its connection to the Baltic Sea, as the land rose in a slow reflex from under the pressure of an age of ice. For millennia until the early decades of the twentieth century it was one of the most important wetlands in Scandinavia, recognised over a hundred years ago for its crucial importance as a migration staging post. It then spent fifty years as a partially drained marsh, until a campaign to get the drainage reversed culminated, between 1988 and 1995, in Sweden's most ambitious habitat restoration project.

I approach the building and see a sign against the glass: 'Closed. Reopening in 2017.' Through the door I see peopleless telescopes pointing through a plate-glass front wall, and through the two layers of glass I see across this southern end of the lake. I walk round the side of the building where I can take in the same view, and on a mound opposite, about a hundred yards away, are 250 cranes, the remnants of the 20,000 who were here at the beginning of the month. In the centre window, a graph shows the pattern of arrival and departure in each year since 2008, jagged multicoloured lines that rise in saw-tooth steps towards their peaks, like a mountaineering chart of some kind. It seems to show that they have arrived earlier, peaked sooner and departed earlier each year. The ones on the mound today seem to be young birds, content to pick away at what remains of the grain that keeps them in place for visitors to view. The centre closed a few days ago, and will remain dormant, like some sullen pupa, for eleven months.

The lake is over 6 miles long, stretching north from this point. In the distance I see what looks like an area of wet meadows and forest on the western shore. I walk north, back along the road towards Härlunda, hoping to find a track that will lead to the lakeside. Initially the road veers away from the lake, and a widening fringe of pasture separates me from the water. The shore is almost a mile away at one point, and then disappears altogether, behind rising ground to my right. I keep walking, and from over this higher ground a huge airborne shape appears. It is a rectangle of wing; the 'flying barn door' coined in 1968 by Geordie Stout, a Fair Isle crofter, and since adopted by field guides as the standard description for the white-tailed eagle. It is an all-dark immature bird, and it flies directly over me about a hundred feet up and over another rise to my left. Immediately, an osprey follows along the same flight path. It is merely buzzard-sized and seems far smaller still, my brain having had no time to recalibrate.

I have seen no track towards the lake, but there is one heading through the trees on my left, in the direction taken by the two raptors. It leads uphill, and I get to the top of the rise hoping the eagle or the osprey might still be in view. There is no sign of either bird, but under the canopy of birch, ash and oak are wood anemones in their hundreds. I have been seeing them everywhere from trains, and now that I am among them, I find that their white starring of the forest floor is subtly accented by small numbers of blue anemones (more often called liverwort in English, *Hepatica nobilis*) and a yellow flower like a tiny lily. It is *Gagea minima*, a delicate relative of the yellow star-of-Bethlehem, with six narrow yellow petals forming a shape that reminds me of a beacon basket, an open inverted cone. I watch a nuthatch peel away a token of birch bark. The bird is a frosty version of the nuthatches I have been seeing in France and at home.

In Scandinavia, the subspecies is whiter, entirely lacking the pink-buff flush on its underside, and with its wings and back a distinctly cooler blue.

North again, into a threshing wind, and finally, a farm track leads right, where a brown sign points to '*Ore Backar 2km*'.

I follow this track and let it twist me for 2 miles among farm fields and through forest, past a birch-wood pasture where a tree pipit sings; past a pasqueflower like a duo of blue-dressed aunts; and nearly past a two-trunked V-signing birch tree. Had I not been looking, by pure chance, at the very spot where it stood like a broken spur and seen its eye glisten faintly, I would have missed a wryneck. When it realises its subterfuge has failed, it surprises me by walking away – not hopping or shuffling – along a horizontal branch. It tolerates my grounded pursuit of it, and I examine every fletch, bar and vermiculation of its crypsis. It is a moth in bird's clothing, or the converse. It is a woodpecker and a *Jynx*. It was used in witchcraft in ancient Greece and mediaeval England, and its Greek name *iynx* became both its scientific denomination and the English word for a magic spell.

The track connects farmsteads: houses and barns among the trees, all with the same liver-coloured timber facings and white window frames. Most have bird nestboxes on the buildings and trees, also in this local livery. The track turns towards the unseen lake for a while, then away again, past a red deer farm, and suddenly, round to the lake shore and a small, empty car park. A footpath sign points to '*Ore Backar, 0,8 kilometre*' and '*Ore Nabb, 1,7 kilometre*'. There is an information panel in the car park with a section in English that explains:

> Ore Backar is a boulder ridge which is surrounded by poor moss-land to the west and rich wetland (Lake Hornborga) to the east. The walking track will take you to Ore Nabb where a bird tower, and a magnificent view to the north, awaits you.

The path climbs the ridge, which is grassy, undulating and hog-backed, elmed and covered in anemones. I want to know how many, so I find a typical-looking square yard and count: 270. 1.3 million to the acre. So at any given time I am looking at at least a million, at a ratio of four swan-white wood anemones to one blue. The blue ones seem to grow paler as they mature, from violet-indigo to etiolated sky. The white ones nod in the wind, and I wonder if this is why they are called windflower in parts of Britain. The nodding seems at first to be in unison, which I can't explain; but when I focus on a fixed area, I realise it is an illusion, or rather a misperception by a brain that wants to make order out of their chaotic rhythms.

The wind is whipping at the lake's surface and making white-crested waves. There is a scourging rain in the wind and no protection from it once I reach the end of the boulder ridge, which forms a ragged peninsula into the lake. It is an effort to study each blurred and buffeted shape on the water, but two red-necked grebes swim close inshore and spare me the work of finding them. Mute swans, anemone-white, dot the lake. They face into the wind, holding their position, and fold their necks into the shallows to feed. I spend a few moments echo-making: finding a visual resonance between the anemones and the swans. The pattern of their nodding – the flowers' furious, the swans' languorous – is, I decide, merely a reflection of the difference in size.

At dusk I return to the forest track behind the farmhouse for my last evening in Höryda Gård and, for now, Sweden. In the distance, there is a sound like cannon fire; tonight is Valborg – Walpurgis Night – traditionally the time for bonfires, fireworks and carolling to welcome the spring. The farm dog has decided she wants to join me, and as the dusk light fades, her teddy bear colour (which matches her soft face, soft-shaggy pelt and sweet demeanour) is first enhanced by the set sun, then slowly melts away into the darkening trees. The last light in the forest is the faint, false luminosity of white anemones.

Marie, a fellow guest at the farm who has travelled from the north for her son's wedding in nearby Skara, explained to me earlier this evening that the wood anemone is the eagerly awaited first sign of spring, and that small quantities would be picked for table posies the moment they appear. The blue anemone is celebrated in a spring song all children know:

Blåsippan ute i backarna står,
niger och säger att 'nu är det vår'.
Barnen de plockar små sipporna glatt,
rusar sen hem under rop och skratt.

'Mor, nu är våren kommen, mor!
Nu får vi gå utan strumpor och skor!
Blåsippan ute i backarna stå,
har varken skor eller strumpor på!'

Mor i stugan hon säger så:
'Blåsippor aldrig snuva få.
Än få ni gå med strumpor och skor,
än är det vinter kvar,' säger mor.

The blue anemone out on the hill,
curtsies and says, 'Spring is here.'
Children merrily pick the flowers,
and rush home, shouting and laughing.

'Mother! Spring has arrived, Mother!
Now we can go without socks and shoes!
The blue anemone out on the hill,
wears neither sock nor shoe!'

Mother in the cottage says:
'Blue anemones never catch cold.
You must still wear socks and shoes!
It's still winter,' mother says.

Two redwings cannot decide between them which should have the last word. They both have three sweet but simple vibrato notes to offer – short, long, long. A woodcock presses its silhouette against the still-lit sky and gives his grunt-and-sneeze 'roding' call. Rain lands lightly. A roe deer barks. A tawny owl halloos.

17 May, Bolletjärn, Gävleborg. 61° 22' N. Evening. Bolletjärn appears below me as a reflective surface glimpsed between densely packed pine trees. As I drop down towards the shore, its waters appear almost black, but at the same time are a light source, reflecting the grey sky in the deep obscurity of the forest. It is a glossed black, simultaneously dark and bright. It starts to pock under a scatter of raindrops, which turns into a steady shower. Under the trees, I am barely aware of any rain, but with every breath of any gust, I hear secondary showers in the rocks and shrubs around me. I reach the shore, watching my step, thinking about words. *Sjärn*, tarn. Diminutive of *sjön*, lake, as in Hornborgasjön. In Northern English we still use a version of the Old Norse word for a small lake, but turned to the Old French *lac* for a full-sized one.

A pair of goldeneye swim away from the shore as I appear, moving 30 yards or so closer to the centre of the *sjärn* but staying within an arc of their own on this side of the water. Across on the far shore another pair are sticking to their stretch of water. I move back a few steps and watch. The rain is now heavy and I hope it doesn't inhibit their courtship display as the snow did a few weeks ago at Loch Kinord.

The pair on the far side swim apart. The male, pied black over white, stands out against the shadows and the edge vegetation. The female is about 30 yards to the right of him, her shades of brown cooling from her mahogany head to her oak back; she dives much. The nearer birds are swimming side by side, and are close enough for me to see a greenish sheen to the male's head as he turns. He has his white cheek-spot and egg-yolk yellow eyes; her eyes are the colour of Cornish ice-cream. Her only bright feature is the yellow tip to her bill. Suddenly, they are paddling across the surface, half-flying, half-running. What unintended movement or noise did I make? Their wings make a rhythmic whistling sound that seems to bounce off the water's surface; the ricochet is like an invisible other pair flying in the distance. As they gain height, the echo-whistle fades to nothing, leaving the pure original sound to circle the arena as they overfly their half of the tarn, before returning with a loud, spray-making skid across the surface.

They resume their side-by-side swimming, diving synchronously, surfacing a few yards away, but no farther apart. I realise that the wing-whistling flight is all part of the act when they take to the air again, unprovoked. Suddenly, the second female is with them, and the three birds fly together for a few seconds, before she returns to her side of the water. When they land, the male of the nearer pair stretches his neck diagonally forward, then tosses his head back as far as he can, till it touches his rump, lifting his chest clear of the water. With his bill pointing skywards for a brief moment, he utters a short burst of sound like the ratchet in a fishing reel. The whole movement lasts about a second. The other female flies towards this near-side pair, and the nearer female takes off too, to head her rival off. She utters a soft, repeated double croak, like a quiet version of a raven's cronking call, and when they land, all four birds have become widely separated. The water is now seething under the rain, and the brief show is over.

It is about nine o'clock, and it will be dark in another two hours, so I move away from the lake to begin my walk back down Stora Bolleberget mountain to the hotel at its foot. When I look up I see a dark-coloured animal about 100 yards to my right, sitting on the path at the top of a rise in the ground. It is about the size of a large dog, blackish and shaggy. I lift my binoculars to my face, but I have carelessly allowed the rain to fall onto the lenses and the image of the animal is blurred. I wipe the lenses with my thumb.

The chances of happening upon a bear are slim, and I know that a lone adult bear would present little danger. I have been walking with a faint hope that I might glimpse one, and a faint fear that one might glimpse me. Twice I have seen large, dark, bear-like shapes in the forest. The first was a rock; the second, more convincing, turned out to be the root plate of a toppled spruce. Now, this smaller shape is certainly animal, and it is staring at me. In the fraction of a second it takes to raise my binoculars a second time, as the animal's bear-ears and pale muzzle sharpen against the optical fuzz of the forest, it strikes me that seeing a bear cub is a more fearful thing than seeing an adult. Almost all cases of attack by bears on humans are by mothers with cubs; if you can see the mother, you are likely to be safe. If you can see only the cub, the mother could be anywhere, and the worst place to be is between the two.

All this goes through my mind in a quarter of a second. For another quarter-second I see the unmistakable image of a bear cub. Then I see the unmistakable image of a remarkably bear-like dog, and a Jack Russell arriving alongside, followed by the heads, the bodies and the legs of two people appearing from over the rise.

Suddenly the rain stops. In a birch tree to my left a wood warbler trills. It may have been singing before, and I notice how similar in timbre it is to that of the forest rain, and how well blended the two sounds would be. The song is echoed by a sialoquent breeze that conjures more water music from the trees in a wave that swells from behind me and fades as it moves ahead. As the cloud cover thins, revealing a hint of late-evening sun from the left, another wood warbler sings, and in a few hundred yards of gently down-sloping path, I cross four territories in all. I see a movement in the corner of my eye, 30 yards to my right. A cow moose ignores me as she trots up the hill towards Bolletjärn. Her footfalls make little sound, and her horse-sized bulk passes lithely between saplings. Her gait is not elegant, but it has a surprisingly light and airy rhythm.

As dusk falls, redwings sing along my route, simple, loud, mellow phrases from the treetops. It is a northern dusk, light enough to find a path, and the sneezy vespertine call of a woodcock alerts me to a stout figure patrolling his patch of night sky.

18 May, Stora Öjungen, Gävleborg. 61° 13' N. This morning I walked back up the mountain and through the forest to Bolletjärn for another look at the goldeneyes. The rain eased overnight and the forest was cool and humid. Long streaks of spruce pollen marked strandlines on the sandy track: pale chalky green braids one or two inches wide, up to 20 yards long. It is a rare pastel, this colour, for which there is no easy combination of words in English.

There was a third pair of goldeneye on the *sjärn* today, and their arrival appeared to have upset the equilibrium. The three males held their stations, one close to the point where the path brought me to the shore, the other two equidistant around the tarn's perimeter. There was no courtship this morning. The females were agitated by each other's presence. Two of them were swimming within 30 yards of each other, while the third kept a warier distance. If they drifted or swam too close, one would skim the surface in a flying run towards the other. Once or twice this led to both birds taking flight, circling the tarn with the soft croaking calls I noticed last night and a high-pitched fippling of air through their wings.

I was delivered back to Bollnäs, where I saw a crane: it was on the town's coat of arms, a white crane with one of its red legs raised, a red sphere held in its claws. I noticed it on a waste bin at the train station, where I waited for Håkan Vargas to take me to see bears.

It is a twenty-minute drive to Håkan's house by Stora Öjungen, a 600-acre lake at the end of a long track through the spruce forest of Gävleborg. En route, we stop at a pet superstore to buy a 25-kilogram sack of oats, which Håkan needs to make porridge for the bears. We arrive, and I am greeted by Håkan's partner Ewa and a young woman, Frida Hermansson. Frida was the first of the night's guests to arrive; she is a graphic designer and wildlife photographer, and this will be her second attempt to see and photograph a bear. A party of four others is delayed, and while we wait, Håkan invites us to take a look at last night's film footage from the feeding area.

'I know we had a bear last night,' says Håkan, 'but I haven't had time to check the camera yet.' He inserts an SD card into the computer, and clicks on the first file. It is twenty seconds of infrared video footage of a large adult brown bear.

'It's a male. He lives just over that hill.' He points out of the window to the high ground on the opposite side of the lake. 'He's been coming quite often in the last few weeks.'

'What time did he come last night?' I ask. Håkan checks the file data – 11.45 p.m.

'Is that typical?'

'There's no typical. He could come at any time after dark, or even before, or not at all. We could see a female instead, or nothing.'

'What are our chances?' asks Frida.

'Quite good, but nothing is guaranteed. I haven't seen a female for some days, and the chances of seeing a male is more dependent on females being around than food, at this time of year. He could be many kilometres away tonight. Ah, but look, we had a second visit, at three forty-five this morning. This one's smaller, maybe a female.'

Håkan leaves for the feeding site, armed with oats, donuts and cinnamon buns. He returns half an hour later just ahead of another car, which parks by the lake. Four people get out. Phillip Lindskog and Elisabet Anjou-Lindskog, and Bertil and Britt Samuelson have driven from Stockholm. The two couples have been friends for years; once or twice a year they spend a few days together, doing something interesting, taking turns to choose what.

'Our last trip was in the autumn, to photograph migrating whooper swans,' explains Phillip. 'It was my turn to choose this time, so here we are.'

'Do you always go looking for wildlife?' I ask.

'No, it could be anything, but we all love nature, so it's a safe choice.'

I have been finding Swedish difficult to follow, but everyone speaks English well, especially Phillip, whose accent is particularly interesting.

'Have you spent time in South Africa?'

'I was born there. My father emigrated from Sweden, and I grew up speaking English. We hated apartheid, and it was getting worse, so my father brought us back here. I'd never been to Sweden until we came to live here, when I was sixteen. I only really learnt Swedish when I joined the army. Even today people tell me I speak "army Swedish", after sixty years.'

We are having lunch at a long wooden table in a wood-panelled room, in a timber-built chalet-style house, which is perched on the hillside about 50 feet above the lake shore. On the cross-beam above our heads are two wolves, the most lifelike taxidermy I have ever seen. They were a pair; shot last year a few miles away, they look down forgivingly upon us.

'It is illegal to shoot wolves in this county, but you can get special permission to deal with troublesome ones,' explains Håkan. 'These two killed five sheep. We tried to stop them being shot, but failed.'

In other parts of Sweden wolves have been culled more systematically, and Håkan explains that most of the decisions about levels of protection, allocation of licences and quotas are decided at county level.

'Last year seven bears were shot in this area', he says.

'Bears? You mean wolves?'

'No, bears. The quota for the county is thirty-seven this year. And we are on the border with Dalarna County, so our bears could be killed as part of Gävleborg's quota, or they could walk into Dalarna and be killed there.'

I am astonished and dismayed. My stomach churns. 'But I thought they were strictly protected under European law! Who ensures the "exceptional circumstances" rule applies?'

'Look, there are hunters in all the county and national hunting administrations who just want to allow bear hunting. There is a special breed of dog, called Plott, which specialises in bear hunting. The hunters say that they must use the dogs or they will go extinct. They tell me "Thank you Håkan, for feeding the bears for us to shoot." Every year I write to the county pleading with them to reduce the quota, but they never reply.'

Two Opel Frontera four-wheel drive cars are parked down by the lake. Phillip has offered to drive one, and takes Frida and me; we follow Håkan, who is driving Elisabet,

Britt and Bertil. We drive for about ten minutes along forest tracks and park the cars, then Håkan leads us silently along a narrow path through birch and spruce. The heathery peat is springy underfoot, and there are short stretches of boardwalk across wet bog. We are carrying our provisions for a sixteen-hour stay in a hide: food provided by Ewa, cameras and binoculars, warm clothes.

Håkan stops and points to a spruce trunk about 9 inches in diameter. There are short, vertical scars in the bark and bluish congealed resin that has bled from the scars, like candle wax. We are avoiding speaking, so he makes a clawing gesture with his hand, tracing the marks with his fingertips. On the ground nearby, embossed into the peat, is a fresh bear print, pointing away from the direction of the hide. After twenty minutes, we arrive at a wooden hut covered in scrim netting. Inside there are seven beds supplied with blankets, each one opposite a comfortable swivel chair by a window slot a few inches wide. Below each slot is a camera hole screened by canvas with a drawstring to secure it round a camera lens. We each choose our station, the few square feet that we will occupy from now – it is four in the afternoon – until eight o'clock tomorrow morning. Håkan advises us to get some sleep, as we are unlikely to see any bears before nightfall, but he will wake us if one turns up. I remove my boots and slip under the blankets.

I wake at eight in the evening, and get up to sit at my window slot. The hide looks across a clearing about 35 yards in diameter. A mist has formed, and two ravens are sitting on a small dead pine in the middle of the clearing. A hooded crow is on the ground, trying to reach the food Håkan has placed in a wood-framed cage. The ravens and crow can reach far enough through the wide steel mesh for small quantities of oats, but can't open the cage, whose frame is designed to collapse when an animal the size of a bear or wolf tries to open it. At nine o'clock Håkan makes tea.

Coffee is not allowed, as it has too strong a smell. I know from my walk last night that there will be twilight until about 11 p.m., although the surrounding trees and the mist will hasten the dark. The next hour is silent and still. The ravens stay in the clearing, cronking to each other now and then. A jay arrives at the food cage, then a great spotted woodpecker. Eventually there is no further movement from the birds, who have finally settled to roost for the few hours of dark to come. I return to bed at ten, and rise again at midnight.

19 May, Stora Öjungen. The mist has thickened, but my eyes have adjusted to the dark, and the shapes of the trees and shrubs are still discernible, as would a bear be should it come into view.

By 3 a.m. the twilight has returned, and I hear muffled chaffinch song from outside. I return to bed, leaving others to keep watch. At 5.30 I rise again; it is daylight. An hour later, Håkan declares the enterprise a failure.

'No bears. Shit.'

The rest of us are silent. We have agreed that we must stay here until eight; only then can we be sure no bears will come, although no one has any expectation of that now. We stay to minimise the risk of disturbance to the bears, rather than to increase our chances of seeing them.

When we arrive back at the lake, the mood has improved. We console each other with reminders that it's what wildlife watching is all about, it's why we do it; the unpredictability, the anticipation. There will be other opportunities, we tell ourselves and each other. Towards the far right-hand end of the lake there is a pair of black-throated divers. They announce their presence with a far-carrying yodel, like a high-pitched wolf-howl. I scan with my binoculars, and catch sight of the two splashes kicked up by their dive. The call continues for a couple of seconds after they disappear, the time it takes for the sound to travel this far. It is as if the call has been released of its bond to the birds, to live a fleeting life of independence.

Yesterday, as I was preparing for a night in the bear hide, another young female golden eagle disappeared in the Monadhliath Mountains, at the edge of the Cairngorms. She had fledged from a nest in Galloway in 2015, one of only two young to have flown from the tiny south-Scotland eagle population last year. Dave Anderson from the Scottish Raptor Study Group satellite-tagged her in July, and followed every detail of her movements on his laptop as she embarked on a journey northwards to Cairngorms National Park. Over the last few weeks he has been looking forward to seeing if she would head south again. However, the tag stopped transmitting abruptly yesterday morning, just after her first birthday. Since 2011, six tagged golden eagles have disappeared suddenly in the Monadhliaths,[2] one of the most intensively managed game shooting areas in Scotland. How many untagged raptors meet their doom there is known only by their slayers.

20 May, Ume River delta, Västerbotten. 63° 45' N. I am walking along a forest track past Villanäs Koloni, a village that has sprung up in the last few years, comprising wood-faced chalets of all colours. A man on a white ATV approaches and stops; he has a stern, roughneck look and I wonder if I am about to be told off for inadvertently trespassing.

'Do you speak English?' he asks. As usual, I have somehow managed, without saying a word, to appear foreign.

'Yes.'

'Are you watching birds?'

'Yes.'

'You must beware. We had a bear on the boardwalk a couple of days ago. So look up, OK?'

I have been following signs for *fågeltorn* – bird tower – from the main port road out of Umeå. The land I am on was deposited as silt in the delta about 200 years ago, the time it takes for a spruce forest to mature, an information panel says. 100,000 tons of sediment is washed down from the mountains by the Ume and Vindel, forming around 10 acres of sandbanks each year. At the same time, the land is rising by 1cm a year, still rebounding from the pressure of the ice that receded at the end of the ice age. The sandbanks are colonised by marsh and meadow plants, then willows and alders, until the land is high and dry enough for spruce.

Before I turn off the track along a path into the trees, I look up, alerted by the call of a crane, high to the north. I find it with binoculars, but my attention is diverted by a bigger shape – a white-tailed eagle with the fully white tail acquired in adulthood after about six years. It has found a thermal and, like the crane, has quickly gained enough height to move on; it soars to the left, in the direction of the delta, where I am heading too. There is another bird in the thermal with the neat outline of a hawk. It has the distinct tough, compact look of a male goshawk, robust but smaller than the more buzzard-like female. I enter the forest, where the dry path rises onto a boardwalk, marking a change to birch-dominated woodland.

Cranes are calling, invisible in the hidden marsh beyond the woods. As the boardwalk bears right the birches give way to willows, and I stop to watch a willow warbler performing a strange display. It is hunched, looking downwards, flapping its wings half-open; it changes posture and stands erect, looking upwards, with its wings slowly flapping in an angelic position. It sings a slow, mournful version of its usual airy song. Never seen that before.

The last dry, sunlit ground is patrolled by a green hairstreak butterfly among the bog myrtles. Then the boardwalk crosses open marsh, with mare's tail and sedges emergent, and marsh marigolds, their lacquered yellow sepals flaring suddenly as their angle to the sun changes with each step. I climb to the top of the bird tower, which is built on a small willow-clad island, tall enough to afford a view to the south-west over the surrounding trees, and from which I can begin to understand the geography of the place. Umeälvens delta, the delta of the Ume, stretches across 8,400 acres downstream of the city of Umeå and is part of the Natura 2000 network. Looking downriver, there are wooded islands and peninsulas and an open brackish bay. I see the steaming chimneys of the Svenska Cellulosa pulp mill at Obbola, 3 miles away on the right bank; from Obbola the E12 road bridge runs across the mouth of the river. An industrial building the size and shape of an English cathedral rises above the trees on the left bank. Beyond the bridge, I can make out the port of Holmsund, where I will go this evening to cross the Gulf of Bothnia into Finland.

There is a channel of the delta in front of the tower, 300 yards wide; on the other side is a willow island which is about 800 yards long, part of a chain that forms a broken peninsula between the arms of the delta. A squadron of little gulls is flying into the wind, each bird dipping every few seconds to pick insects off the brackish surface. They fly only marginally faster than the headwind, giving them the ground speed of a slow walk; to dip, they stall, and pick their prey with the tips of their bills, rising to turn downwind and upstream to begin the search again. There are twelve of them, and the amplitude and rhythm of their movement makes of their separate flight lines a kind of collective machine. They never tire me of their balletics; I first fell in love with the smallest gull two years ago, when I watched an immature bird at Titchwell in Norfolk. It was a balmy evening in early April, and insects were rising. The bird had a distinct way of dealing with flying prey, more swallow than gull, playing the faintest breeze to gain advantage over the dizzy new midges.

To the left, 600 yards downriver, are two more islands in the chain, both low, grassy and treeless. The crane calls I have been hearing emanate from there. I count seventy-six, scattered in irregular groups along the islands. Sitting apart is an immature white-tailed eagle, like a freshly pollarded stump. Farther round to the left the water has an intense silver glare; the wind has stippled the surface and turned it into a glistering fabric that is painful and irresistible to watch. Two hundred or more whooper swans are camouflaged in the smithereens of sunlight. They call continually, an under-murmur to the cranes' split-note fanfares. I am reminded of the comparison Sibelius made in his diary after he witnessed the spring return of the swans in April 1915:

> Their call the same woodwind type as that of cranes, but without tremolo. The swan-call closer to the trumpet.

There is also a call that is like something out of Norse mythology. A tortured cry, loud and water-borne, from the white-tailed eagle. I have heard it four times this morning, and the curlew twice. These four disembodied voices are the music of the delta.

The whooper swan is called *sångsvan*, 'song swan', in Swedish. The *carmen cygni*, the mournful music uttered by the dying swan, is one of the most misapprehended avian legends, usually unbelieved, but in fact true. The whooper swan has a long, looped trachea that enables it to send deep, resonating calls through the mist, keeping the flock in close contact over long distances. Occasionally, when a hunted bird is collected in the last moments of life, a long, final exhalation produces a stream of pure, flute-like notes. Living birds have simple, less fluty but no less resonant calls, which they occasionally synchronise, sending waves of homophony through the group. The delta flock sends a choral rippling that reaches me against the river's flow. The cranes' solid and irregular cries are like sonic islands mid-stream, while the faerie-song of curlews is from the swirling realm of the air, like a sound originating from a distant source and passing overhead. The Sámi concept of realms – of the living and the dead, the underworld, the middle earth and the spirit world; their denizens and tutelars, the shamans who journeyed between them, the birds and animals that acted as guides – may owe something to soundscapes like this. If so, the white-tailed eagle is the bird with the voice to communicate across the divides, haunting all realms.

Finland

21 May, Liminganlahti, Northern Ostrobothnia. 64° 49' N. I arrived in a small Finnish town at midnight, with a heavy pack and a bus to catch at 2.30 in the morning.

In O'Malley's on Hovioikeudenpuistikko, across the street from Vaasa bus station, an American pianist called Barrelhouse Chuck* and his band were playing twelve-bar blues until two. I recognised the faces of some fellow migrants from the MS *Wasa Express*, which set sail from Umeå at six o'clock Swedish time last night, losing an hour of clock-time halfway across the Gulf of Bothnia.

I came south and east to the port of Vaasa in order to continue north by bus. Until then, my tee-total trip had been made easier by Sweden's punitive alcohol tax and hilarious booze-retailing system. The latter is a state-run monopoly, to be found at the green sign of the Systembolaget, in which liquor is dispensed with solemnity in the atmosphere of a high-street pharmacy. In O'Malley's I tried to make a pint of Guinness last by reviewing my notes of the previous few hours, and adding contemporaneous ones:

> *A selection of middle-aged blues fans: men with grey beards and short-cropped hair, or beards and long hair; women with smokers' voices. There are hard drinkers, steady drinkers, and Swedish alcohol tourists. Plus a lot of well-dressed youngsters.*

A young golf professional (I would have guessed if he hadn't told me: *immaculate polo shirt, white Nike trousers, baseball cap and sunglasses facing backwards. Fresh-faced 'all-American' look*) called Wili asked me in English if I spoke Finnish, and when

* Died 12 December 2016.

I said no, asked me if I spoke English. He asked me what I was writing and I explained that I had been following the progress of spring through Europe, and that I was tidying my notes while waiting for the bus.

'That's a cool project,' he said. I asked if he was from Vaasa. No, he was from Kokkola. He was in Vaasa for a tournament, and had a practice round tomorrow, which I told him was today.

'Yeah. I'd better go.' And he did.

I had a second pint which was two-thirds finished when I decided to leave it, and O'Malley's, and make my way to the bus station by a roundabout route to kill time. I came to a square with a large church on one side and enclosing a few dozen large, widely spaced trees. The sky appeared dark and the lights of the town were bright enough for a chorus of fieldfares to sing by. They were too bright to allow a man to pee in obscurity behind one of the trees, I noted, but as everyone was down at O'Malleys, I figured I wouldn't be noticed.

The fieldfares were singing at the bus station too, and I sat to watch the sky grow a fraction lighter to a colour I don't remember ever seeing before – something like the faded but still deep blue of antique willow pattern, washed with the faintest taste of purple. Somewhere on the bus journey to Oulu the dawn broke, and I knew I would see no more darkness for twelve days.

Twenty miles south of Oulu the bus passed the turning to this bay, and might have dropped me off if the rules had allowed. I was lying across two offset pairs of seats, impaled by seat-belt fittings; I lifted my head and twisted to look out of the bottom few inches of the window, just as the bus flushed a short-eared owl from a roadside fence post. It was lit by an amber light that emphasised its easy flight and reflected off its amber eyes.

Swifts have arrived in Oulu, three of them screaming between the buildings that line the streets around the bus station. I take a taxi to Virkkula on Liminka Bay (Liminganlahti in Finnish) rather than wait four hours for the Saturday shoppers' bus back to Liminka. Liminganlahti is Finland's largest wetland, and like all national parks and most major nature reserves, as well as all Finland's state forests (in total over 35 per cent of the country), it is managed by Metsähallitus, a state-owned enterprise. There is a guesthouse there and I have a room booked, but I arrive before the visitor centre opens, so I take my backpack and day bag along the boardwalk and up the birdwatching tower, where I scan a corner of its 30,000 acres and 70 miles of shoreline. Behind me to the south, a croissant-shaped fringe of willows separates the water from the land. In front, a 5-mile lobe of water within a bay twice as wide, itself an inlet off the north-east corner of the Gulf of Bothnia. I look over the edge of the platform into a straight channel cut into the bay. Beneath the surface there are water lily leaves, submerged by the morning tide. Water lilies in the sea; a freshwater sea. The Gulf has its outlet 450 miles to the south, into the Baltic. The Baltic itself is merely brackish. It is a broad gulf off the North Sea, whose salinity is five times that of the Baltic, and ten times that of this bay.

Scattered through this hemisphere of water and sky are some of the birds that have become as guides and counsellors on my journeys. Eight cranes are feeding quietly; there are about 150 whooper swans, and I see two solid, sculptural white-tailed eagles on the ground in the distance. A black-tailed godwit is displaying over the wet fields, uttering the same sharp three syllables I remember from the Ouse Washes. A ringing, piping, two-note call alerts me to a chase: a greenshank is power-sculling the air towards me, keeping low to the ground, matching its target's jinks and banks. The female sparrowhawk takes refuge in a tussock 30 yards to my right, and the greenshank disappears, merging into the marsh grass. The sparrowhawk makes three air-gulping wing flaps to become airborne again, only for the greenshank to appear out of nowhere. Reaching fighting speed before the hawk, she delivers a calculated near-miss that almost sends the hawk sprawling back to ground. Although she lacks the hawk's weaponry, she must escort it from her airspace. Hundreds of yards away, on a grassy spit, there are about thirty ruffs in summer plumage. Despite the distance, I am pleased to see them.

I have existed at a perpetually incipient spring as I travel north, noting signs and hints, reading runes. Tomorrow I will arrive in the land where traditionally there are eight seasons, called by the Sámi: *Čakčadálvi* (early winter, a time of wandering), *Dálvi* (winter, a time of caring), *Giđđadálvi* (early spring, a time of awakening), *Giđđa* (spring, a time of returning), *Giđđageassi* (early summer, a time of growth), *Geassi* (summer, a time of contemplation), *Čakčageassi* (early autumn, a time of harvesting) and *Čakča* (autumn, a time of change). The male ruff is the epitome of *Giđđa*, of spring fulfilled: concerned only with finery, sex and fighting. The female, too, through her collusion. This –the male – is the bird that changes his appearance each year from the standard sandpiper template to something resembling an Elizabethan court dandy. He acquires an extravagant collar of long, foppish feathering at his throat and at the back of his head, available in a choice of colours, black, brown and white being the most common, barred or plain. The choice is made for him by his DNA, and he will adopt the same pattern each spring of his life. The female keeps her cryptic coloration throughout the year, and there is a significant size difference between the sexes, on account of which the female has traditionally been given the name 'reeve' in English. In old bird books, the species was known by a unique composite name: 'ruff and reeve'.

In 1676, Francis Willughby's *Ornithology* reported that:

> at their first coming in the beginning of the Spring there are many more Cocks than Hens, but … they never cease fighting till there be so many Cocks killed, as to make the number of both Sexes equal.[1]

In reality, their legendary fighting is rarely to the death, but is instead staged like a Tudor joust, with just as much intrigue at the sidelines. The sexual politics of the species are no less complex than ours, with salacious details continuing to emerge from recent research. Some males – the darker plumaged ones – succeed by out-competing the rest in glamour and aggression. Others, the smaller white-maned birds, are non-territorial but adept at stealing mates from their more aggressive rivals. They are tolerated in the court of the dominant males, as a large coterie attracts more females. In 2006 the first known case of female impersonation in birds was reported, when it was discovered that about 1 per cent of male ruffs have a female-like appearance. It seems that not all reeves are on the lookout for rough machismo. Plain, unadorned males hang around the fight arenas, known as 'leks', where dark-feathered males go all-out to attract mates and pale-feathered birds play a subtler game. The rare cross-dressers also steal mates, but by gradual insinuation. By lacking ornamental feathers and mimicking females, they are able to hide in plain sight before making their move. Earlier this year Professor Terry Burke from the University of Sheffield discovered a so-called supergene in the ruff that has allowed several traits – including hormones, feathering, colour and size – to evolve together and create these parallel behaviour patterns.

At 10 a.m. I check in at the centre and study a large, wall-mounted map of the bay. There is another birdwatching tower, at Puhkiavanperä, that seems to be within walking distance, and I return to the road to head west round the bay. The road passes through fields and among small farmsteads, a walk of limited visual appeal, but zesty with skylark, yellowhammer and whinchat song. After two miles a track takes 90-degree turns around square fields before crossing the Lumijoki river, with its small jetties and fishing platforms every 30 yards or so along the birch-wooded left bank, and open fields adjacent to its marsh-fringed right bank.

After another one and a half miles I come to a small parking area with a boardwalk leading off it for a final half mile into the reed bed that fringes the bay. At four miles, the distance to the second tower is farther than I had anticipated. I have time only for a cursory scan of the reeds: black-headed male reed buntings dot the reed tips like seed-pods. Sedge warblers remain hidden, but their free improvisations suffuse the reeds from all directions. Two cranes call as they fly low over me and land close by, but out of sight, behind the reed bed. The return walk will need to be something of a route march, for I have arranged to be back at the centre by two o'clock for a conversation about *Kalevala*.

I arrive back at the centre, into a wide entrance lobby. Around the walls of the lobby is a series of watercolours, all depicting owls, the work of artist and broadcaster Minna Pyykkö, who has arrived from Helsinki to present an exhibition of her work. Minna is the face and voice of Finnish nature on television and radio, regularly hosting the *Areena* programme and *Minna Pyykkö's World*. When I arrive I see a small figure in a peach-coloured sweater with long brown hair. She is speaking to a middle-aged couple, and I wander along the line of paintings while I wait for her to become free.

The pictures are all between about 10 and 18 inches square. They are all of owls, of the ten native Finnish species. Some are portraits – studies of attitudes and moods that among birds, only owls can betray. These are finely and accurately observed, yet rendered with an economy of brushwork. Others are landscapes and lightscapes with the owl a discreet luminous presence. In one painting, the sky is the colour of the one I watched from Vaasa bus station last night. There is a human figure, a girl, and an owl slips into the darkness unnoticed. Another is of a bird ringer, half lost in the gloomy landscape, holding an owlet on its back against his knee.

'When I saw him ringing the owl,' Minna tells me, 'it reminded me of Hamlet holding the skull.'

'Alas, poor Yorick?'

'Yes, the delicate way he handles it and all his attention focussed on one spot.'

I suggest to Minna that there seems to be a quality in her work that places it at the boundary between reality and fantasy. She leads me to one showing a figure of a youngish, casually dressed man in the middleground, looking out of the picture. In the foreground is a great grey owl flying towards the viewer.

'That's my father, who died many years ago when I was young. The owl is a kind of spirit.'

Like generations of Finnish artists, Minna has created many works inspired by *Kalevala*, Finland's national epic poem. As epics go, it is relatively recent, completed in 1849 by Elias Lönnrot, a physician who used his spare time to collect and compile the ancient oral folk tales and myths of the Finnish people and then used them as the raw material for his masterpiece. I wanted to know whether *Kalevala* reflected some deep-rooted connection between Finnish people and nature.

'Birds play a big role in *Kalevala*,' says Minna. 'From the very beginning – the world is created out of the shards of a goldeneye's egg. In *Kalevala* birds are companions to people, they whisper advice, they tease, they lift and carry tired travellers, they attack and fight with people. They also feel cold in winter and are happy when spring comes.'

When the original poems were spoken, the earth was believed to be flat. At the edges of earth was *Lintukoto*, 'the home of the birds', a warm region in which birds lived during the winter. The Milky Way is *Linnunrata*, 'the path of the birds', the route the birds took on their journeys to *Lintukoto* and back. In modern Finnish, *lintukoto* means a safe haven, an imaginary happy, warm and peaceful paradise.

'Birds bring comfort and joy to people,' Minna continues. 'They bring a person's soul to him at the moment of birth, and take it away at the gates of death by the river of Tuonela.'

'Tuonela. I know that name. Sibelius's *Swan of Tuonela*.' It's a favourite of mine: a long, sinuous and other-worldly melody on the cor anglais that lasts for eight minutes with no theme repeated, like the waters of a slow river that can never turn back on itself.

'The hero Lemminkäinen tries to kill the swan, but gets killed himself.' Minna explains. 'I have been many times to the archipelagos in the Baltic to paint a duck … we call it *alli* … what is it in English?' She calls over to a woman standing nearby. 'What's *alli* in English?'

Ulla Matturi, the centre manager, joins us. '*Alli*? It's, er … long-tailed duck, I think. Yes, long-tailed duck.'

'That's it,' says Minna. 'It's the most frequently mentioned bird in *Kalevala*. We get millions of them in the Baltic, and then they fly overland to breed in the Arctic. I paint them in flight, in great flocks. They are like a single organism, a driving force, when they move like that. We call them the Bringers of Spring.'

'It's late for them now, but you should see some further north,' says Ulla.

'As for Finnish people being connected to nature, that's changing. Children's roaming range has been cut right down.'

'It's the same in Britain,' I say. 'Studies show that children cover only 10 per cent of the ground they used to.'

'Tell him about the million nests campaign,' says Ulla.

Minna tells me that she and her husband the writer Juha Laaksonen had an idea for a campaign 'to give Finland a centenary gift of birdsong' by installing a million new nestboxes in the country's gardens, parks and forests in time for spring 2017. The Finnish lands were part of the Kingdom of Sweden between the twelfth century and 1809, when it became an autonomous Grand Duchy of Czarist Russia. In December 1917, when Russia was in the grip of revolution, the Finnish nationalist movement, which had been gaining momentum with the appearance of *Kalevala* and the music of Sibelius, seized the opportunity to move for full independence.

'We launched the idea in the autumn to give ourselves two springs to reach our target. Already we have 750,000 boxes installed.'*

'The public are encouraged to make or buy a nestbox, and take selfies showing the box in place,' explains Ulla. 'There is a website for uploading the pictures, which is how Minna and Juha are able to monitor progress.'

'Finland has more forest cover than most other countries, but it is almost all commercial, and we have lost vast numbers of old trees. Birds have fewer natural holes in which to make their nests. Studies show that 1,000 species of animal and plant may be lost from Finland because of the decline in quality of our forests.'

Finland has its own Forest Certification System, choosing to shun the internationally recognised Forest Stewardship Council certification scheme. A report entitled *Certifying*

* The target was reached on 22 February 2017.

Extinction? by Greenpeace, the Finnish Association for Nature Conservation and the Finnish Nature League shows that 564 forest species are classified as threatened and 416 as near-threatened in the Finnish Red List, most as a result of modern forest management.[2]

I have travelled through the night, to a place where there is no night, so I decide to sleep a few hours and get up later for a midnight walk. In bed, I read from *Kalevala*, a few of its 22,795 lines.

> Wainamoinen then departed,
> Empty-handed, heavy-hearted,
> Straightway hastened to his country,
> To his home in Kalevala,
> Spake these words upon his journey:
> 'What has happened to the cuckoo,
> Once the cuckoo bringing gladness,
> In the morning, in the evening,
> Often bringing joy at noontide?
> What has stilled the cuckoo's singing,
> What has changed the cuckoo's calling?
> Sorrow must have stilled his singing,
> And compassion changed his calling,
> As I hear him sing no longer,
> For my pleasure in the morning,
> For my happiness at evening.
> Never shall I learn the secret,
> How to live and how to prosper,
> How upon the earth to rest me,
> How upon the seas to wander!
> Only were my ancient mother
> Living on the face of Northland,
> Surely she would well advise me,

What my thought and what my action,
That this cup of grief might pass me,
That this sorrow might escape me,
And this darkened cloud pass over.'[3]

And I read about Northern Ostrobothnia, finding that the crane is the official emblem bird of the region, and to my considerable pleasure, the stoat is the official regional mammal.

22 May, Liminganlahti. At midnight I walk out from my room and down the boardwalk through light rain. The sun set exactly an hour ago; astronomical midnight is an hour away and the sun will rise again at 3.30 a.m. Last night in Vaasa there were two hours of darkness; tonight all three measures of twilight – astronomical, nautical and civil – continue until dawn. The clockwork 'tick-tock' of a snipe seems to emanate from the humid fabric of the air itself, and sets a presto tempo. Two male whinchats out on the marsh lay out a relaxed refrain of short, airy phrases that start scratchily but quietly accumulate bell-like tones that cluster around a central pitch. Overlaying fast and slow tempi was a favourite technique of Sibelius, and the composer Michael Finnissy has described perfectly how Sibelius depicted the Finnish landscape in sound: 'His pieces have a long perspective, always in long-shot, not close-up; then you get close-up events that seem unusually magnified in importance because there's such a huge contrast between vast and close-to.' I return to bed with the thought that Sibelius will have heard these sounds on spring nights, and absorbed them into his art.

I am first into the kitchen, and set to work trying to get some coffee going. I am soon joined by another man, about fifty years old, with short, thinning dark hair. He is carrying a laptop that is streaming news in German, which he places on the long, communal table.

'Coffee?'

'Is there enough for me, too?' he asks in an accent I can't quite place, but which doesn't sound German.

'Can be.' I sort out the coffee while he removes the self-service breakfast items from the fridge. He is Nadko, a Bulgarian living in Bavaria.

'I'm here to see lekking ruff,' he explains, switching off his laptop. 'I flew into Oulu two days ago and hired a car. I spent all yesterday in a hide watching them, and should manage another three hours today before I have to drive back for my plane to Munich.'

'It's a species I'm hoping to get to know on this trip, too. Is it a research project?'

'No, not at all. I'm in the pharmaceutical industry. It's just something I wanted to see and I had a free weekend. I brought my camera but I prefer just watching them, trying to work out what is going on.'

I tell Nadko that a ruff lek is something I've seen once in my life, about forty years ago at the Ouse Washes. He asks me why I have come to Liminganlahti, and I explain about my journeying at the front edge of spring.

'My spring began in Hokkaido in January,' he says. 'I went for the weekend to watch red-crowned cranes dancing. That was amazing too.' I begin to sense we are kindred spirits. 'Then in April I had a weekend in Hungary to see great bustards lekking.'

'So, is that what you're into? Birds that lek and dance? How long have you been interested in that?'

'Two years. My wife is on a three-year assignment in New Jersey. For ten days every month I go out there to be with her. The rest of the time I need something else to do. So I thought I'd try bird photography and watching. It's mainly watching. I made a list of all the things I wanted to see and that's why I'm here.'

The kitchen door opens and a man about the same age as Nadko, but contrastingly pallid and rotund, walks in. Nadko says 'good morning' and he says 'good morning' back to Nadko, then turns to me.

'*Hyvää huomenta. Oletko Suomalainen?*' I recognise the last word.

'No, English.'

Seppo is a postman from Helsinki who has driven 370 miles to find a gull-billed tern which has been reported from the reserve. It is a rarity in Finland, and Seppo explains that he would like to add it to his list of the birds he has seen in his country, but that it is no big deal if he doesn't see it. His former wife and his current girlfriend don't share his enthusiasm for waiting hours for a rare bird to show, although his girlfriend made an exception for an oriental cuckoo that turned up near the Russian border last year.

Three middle-aged men, on what seem to me to be three different missions. Whether the differences would be appreciated by anyone else, I cannot decide.

'What about Leicester?' asks Seppo, and Nadko repeats the question.

'Yeah, what about Leicester?'

Three weeks ago Leicester City F.C. secured the English Premier League title, having started the season with nominal odds of 1,000 to 1. Strange that neither Minna nor Ulla mentioned it yesterday.

The Finnish bus network is efficient and comprehensive, but long-distance services are infrequent. It means I have time for a last visit to the bird tower, but not enough time to join Nadko at the ruff lek 8 miles round the coast. This morning lapwings have taken control of the first 60 feet of airspace over the marsh, loosely sketching the borders of their territories. In the cool morning air, sounds reach across a thousand uncontoured acres, sounds of infinitude. Greenshank, whooper swan, redshank, curlew: calls whose acoustic properties allow them to haunt the air, and which are therefore as much a part of the landscape as the shapes and colours in the marshes and the sky. The murmuring undertow of swan-call that shimmered in the Ume delta two days ago is attenuated with distance here. Greenshank, redshank, ticking snipe, crane, white-tailed eagle: the *cantus arcticus*. I think of the ailing Einojuhani Rautavaara, a musico-ornithologist who has taken Sibelius's mantle as the voice of the Finnish landscape. He must soon pass it on* to a generation that, across the Nordic countries, has remained attuned to and inspired by nature throughout the prolonged urban preoccupation of composers elsewhere in Europe.

The bus leaves Oulu on time at 11.30 a.m. for the three-and-a-half-hour journey to Rovaniemi. Most of the houses and farmsteads are in the same lap-wood style and many painted the same liver-and-white as in Sweden; but not all. There is a pastel tendency here, rare colours like the tide of spruce pollen I saw at Stora Bolleberget, the mint ice-cream green of the forest lichens or the lavender grey of a lightly overcast sky. Carved out of a mainly forested landscape, there are rectangular grass fields with pyramids of white silage bales. There are clear-fell areas where huge hydraulic ploughs have left battlefield scenes. Håkan protested a similar ravaging tendency in Sweden.

* Einojuhani Rautavaara was born in 1928 in Helsinki and died there on 27 July 2016. *Cantus Arcticus* (1972) is a symphonic concerto for orchestra and his own field recordings of birds.

Rovaniemi, Lappi/Sápmi. 66° 30' N. My collection of cuckoo calls grows as I walk through the narrow park that separates the town from its rivers; an exact major third to add to all the minors and microtonal thirds-and-a-bit. Rovaniemi sits where the Kemijoki river greets the joining Ounasjoki, forming a kind of inland delta before flowing into the northern Gulf of Bothnia at Kemi. I head for the banks of the Ounasjoki, upstream of the confluence. The river must be high, for its waters have extended beyond the fringing willows and crept into the birch; of a dozen 16-foot shallow-draught dories that have been pulled ashore into the trees, several are half in the water. A bird is calling at the river's edge, hidden from view behind a tangle of willows: three fast notes on different pitches, slurred together, with the last one shorter, more staccato, than the others. They are repeated over and over again, with a note's rest before each repeat. I sit on one of the upturned boats to pull on a pair of Feetz folding wellies, possibly the most ridiculous-looking apparel since the codpiece. But they fitted into my coat pocket and now get me into the floodwater so that I can reach the river bank and look for the source of the voice. I hear it again, its owner still hidden, and at close range there is a quiet detail inaudible at more than 10 yards. The three notes are preceded by a very rapid double grace note, a dry, extremely quiet 'ti-tic'. It is as if the two parts to the call are intended for two audiences, one near, one far. Suddenly, the bird reveals itself by flying on stiff, shivered wings across a narrow braid of the river to a low, grassy island 30 yards away. A common sandpiper, followed by another.

As I watch the sandpipers cross the water, a distant movement catches my eye. On the far bank, 400 yards away, a group of about sixteen ruffs have flown in, and immediately start to lek. There seem to be four black-ruffed birds, about twice as many whites, and about six tans. From this distance it is impossible to discern a pattern to the fluttering, jumping, upright posturing, ruff-splaying and bowing, lunging and parrying; the hierarchy, dominance, subordination insubordination and subterfuge. It has been drizzling for much of the afternoon, but the evening sky has cleared and in the low evening sun the lek is like a prematurely ignited firework display in its energy and randomness, and in its short-lived splendour. After less than half a minute, the birds fly upstream and out of sight.

As long as the sun is above the horizon I know it is not yet 11.30 p.m., but otherwise I have no idea of the time: I have left my phone in a hotel on Ounasvaara fell, where I dropped my bags before walking the 3 miles back into town to the river. I can see the hill from here, clad in old-growth forest, with the accoutrements of the ski resort that it becomes during the snow time; it is now largely deserted.

On Ounasvaara fell the night is bright enough (despite the return of the drizzle and its gradual deliquescence into heavy rain) for me to follow a network of marked paths through the forest. The pines are scabrous with acid-loving lichens: *Hypogymnia physodes*, like overlapping lava flows radiating from a point and frozen in place; and *Bryoria fuscescens*, like tangled hair, pearly with rain. Imagine such a forest without them, all plain brown trunks and boughs. Stand still, and their fuzzing of the trees' outlines and colours creates the illusion of two dimensions, like an impressionist painting. Lichens can be overlooked, but their impact on the scene cannot be ignored. Each type represents a marriage between kingdoms, a colony of single-celled plants living among fungal filaments. The fungus cannot live without the alga, and the reverse is sometimes (but not always) true. Only in combination do they take the form we see. A sound shoots through the forest, like a laser of air, as high-pitched and unwavering as a dog-whistle, and provides the depth of field, the third dimension that makes the painting habitable.

It is the call of the hazel grouse, a bird whose camouflage and furtiveness make it unlikely that I will see it. It is a bird of the lichened and leaf-strewn forest floor, with its underside patterned in the shapes and colours of autumn birch and bilberry leaves, and with its lichen-patterned back and wings.

The path rises onto a boardwalk, where glaciation has left depressions in the bedrock to fill and form sphagnum-clogged mires. Redwings inhabit this forest more densely than in any of the places I have been to so far, their songs signalling proximities rather than territories. There are song thrushes too, and robins. A pied flycatcher completes the midnight line-up. The redwing's song is simpler than the rest, but I am discovering how remarkable it is none the less. Along the boardwalk I pass close to eight or nine song-posts, and hear others in the forest to either side of the path. It is as if the song thrush has given each male redwing one phrase from its infinite songbook to learn. And having learnt it, each can only repeat its short gifted song, over and over. Each redwing, then, has its vocal signature – short, sweet and individually monotonous. But as you pick a path from song-post to song-post, there is a communality to their singing; the whole is a part-song. In my improvised myth of explanation, only the song thrush knows it all.

23 May, Arctic Circle, 66° 34' N. 'Last of all is the Scythian parallel, from the Rhiphean hills into Thule: wherein it is day and night for six months by turns.'

Pliny the Elder's *Natural History* of AD 77 explained – in effect – that places on the earth's surface that share the same exact pattern of variation in night and day sit on the same parallel line scored onto the globe. The line that unites all the places where, one night a year, the sun fails to set,[4] is named after the most northerly of the patterns in the night sky: *Arktikos* – 'of the bear'. The bus crosses the line at midday exactly; it is marked by a blue sign and a larger yellow one announcing (in Finnish, Sámi, English and German) that we are entering the reindeer husbandry area that covers the whole of Finland north from here. Then another sign, larger again, for the Santa Claus Village, which occupies a 20-acre site to the right of the road. It looks like the kind of place you take children to when you are desperate to cure them of their belief in Father Christmas. I see my first reindeer within a few minutes – two grazing in a field to the left of the road – and a white one a few miles farther on, in a field to the right.

I am heading for Inari, 204 miles – and, according to the bus timetable, five hours and ten minutes – away. This is Sápmi, the cultural region that spreads across 150,000 square miles of Norway, Sweden, Finland and Russia, the most sparsely populated region of Europe. We cross into Sodankylä district, with its 1.9 inhabitants per square mile. It is boreal forest country – taiga. As we move north, I note that the proportion of birch to conifers seems to be increasing; and that there are more areas of pure mire and wet woodland. We pass areas of virgin mire that have been ploughed for forestry, marked by linear pools of water-filled plough lines. The controversial Finnish Forest Certification System purports to protect such peatlands, but in reality allows drainage in most cases, putting unique biodiversity at risk and turning peatlands from carbon sinks into carbon sources.

The road is single-carriageway, except at a few junctions and settlements where there are filter lanes to siphon traffic off to left or right. It swings along the contours of hills or flows with the sweep of rivers and lakesides; but such terrestrial deviations go unnoticed at map scale: overall, the road runs almost due north between Rovaniemi and Inari. As I sit looking out of the right-hand-side window, it is the first time since beginning these journeys that I have a true sense of a line of longitude sitting under me – roughly 26° 22' E, as easterly as Bucharest. For 55 miles, we follow the Kitinen river upstream to its outflow from the Porttipahta

Reservoir. The farther north, the more the forest thins and is interspersed with increasingly extensive mires – flat and horizon-filling, with wide, shallow pools reflecting the overcast sky. The Finnish term is *aapa*, minerotrophic peatlands as distinct from the nutrient-poor raised bogs of the south.

The bus slows sharply as a reindeer crosses casually in front of us, left to right, and a large lake appears on the right. We stop and a woman gets off, leaving four of us in the bus to continue north. She is in her seventies, and we leave her in the middle of nowhere, where there are only trees and water. I watch as she walks down a track towards the lake shore – her grey hair held back with a comb, with her pale-green windcheater, slacks, print blouse and unmatching print scarf – and wonder, when I notice some houses on the other side of the lake, if she is going to get into a boat. The bus moves on and I'll never know. A few miles later we stop and no one gets on or off, but the driver accepts a white polystyrene fish box from a man in beige overalls who is standing at the side of the road by a white van. '*Nordlaks Hotelli Inari*' is written in blue marker pen on the box, which goes into a cargo hold under the floor.

A sign in Sámi and Finnish announces the border of Inari district. With 1.2 inhabitants per square mile, it is the least densely populated part of the country. A swallow streaks across my view from the bus, the first I have seen since crossing into the Arctic; I wonder if it will be the last, and I wonder if it has flown over any of the places I have passed through to get here. At Ivalo bus station two middle-aged women get off, and one young woman gets on. There is a white minibus packed with Russians on route to Murmansk. The electronic sign-board on the bus is registering 18 degrees Celsius outside, and I get down into the warm afternoon sun. A pied flycatcher is singing, hidden in the bushes behind the bus station, and I hear a willow warbler farther off in a stand of poplars that line the main road. They both have short songs of separate but elided notes that slip down in pitch. The flycatcher's is introspective and rich-toned, half the speed of the willow warbler's fresh and airy motif.

My destination is 25 miles farther on, through a waterscape at the edge of the huge Lake Inari, whose west shore we skirt and whose inlets we cross on causeways. There are scattered pairs of goldeneye in the rocky indentations of its coastline.

Inari, Lappi. 68° 54' N. The faster a river flows, the more it presents an image in stasis. I walk upstream along the right bank of the Juutuanjoki, which is swollen by meltwater and crashes over its boulder-strewn bed, shaping itself into many-braided breaker-waves. The waves themselves stand still: the water may be agitated to the point of opaque whiteness, and each ton or drop of water may speed along the channel without relent, but there is a constancy of fetch and form in the waves; they froth in one place, feather in another. There is no surface-drifting flotsam to draw the eye into movement; even a drake goosander is swimming on the spot, facing upstream, ducking his green head beneath the surface to watch for prey. In the pine trees along the river, where the Hotel Kultahovi has provided a wide feeding table and covered it in sunflower seeds, there is a pair of red squirrels, chasing in skewed spirals around the trunks. One is clothed half in grey, shedding its winter coat. Squirrel skins were used as currency in Finland before the Swedish krona became the adopted coinage in the early 1700s; the modern Finnish word for money – *raha* – is derived from skrahā, a proto-Germanic word that meant 'squirrel pelt'.

The track deviates away from the river through the pine woods, via the perimeter of Inari School, on whose cream-yellow walls its name is written four times: Inarin Koulu, Aanaar Škovlâ, Anára Škuvla and Aanar Skooul; in Finnish, in Inari Sámi (*anarâškielâ*), in Northern Sámi (*davvisámegiella*) and in Skolt Sámi (*sää'mǩiõll*), four distinct and mutually unintelligible languages that converge here. Lake Inari is the centre of the homeland of the roughly 800 people of Inari Sámi heritage, of whom fewer than 300 speak *anarâškielâ*. This distinct group traditionally live by fishing and hunting, the only Sámi whose lives are not inextricable from those of the reindeer.

I have not yet come to treat the pied flycatchers that have graced every part of my journeys since Bolletjärn as the commonplace that they are: their fiscal design is too perfect for me not to stop in each territory I pass through. Their puritanical plumage belies their open-marriage policy. They practice successive polygyny: the males leave their home territory once their first mate has laid her eggs. Males then create a second territory, often over a mile away, and attract a second female to breed. Then they return to the first female, exclusively providing for her and their offspring. To use human language, it seems that some females are able to secure their mates' ultimate attention – as wives – while others, the mistresses, become one-parent families, at least until the legitimate offspring have fledged. The wives have better reproductive success, while the mistresses produce 60 per cent fewer offspring. One theory to explain the apparently disadvantaged role of the mistress is that males who are able to secure two breeding partners sire equally successful male offspring, so that while the mistress may raise fewer children than average, she raises many more grandchildren than she would with a congenitally faithful husband, and secures her genetic legacy that way. This is the 'sexy son hypothesis', first proposed by Rothamsted geneticist Ronald Fisher in 1930.

There are brown trail signs every 50 yards or so as the path enters the forest and climbs up a rocky rise through the pines. I hear a distant, unfamiliar call to my right, like a soft, deep toy trumpet. I walk towards the sound, surprised at how far – about 200 yards – I have to go to find the source. It is a male brambling, a close relative of the chaffinch. He is standing against the light on a large side branch, and I shift position so that I can see him against the trunk: he is in full breeding plumage, with black head and upperparts, pale apricot-orange breast and epaulettes and white belly. Each time he calls he lifts his head to about 20 degrees from the vertical. I return to the track, and after another 200 yards or so another brambling calls. I see it in a half-dead pine, its back to me, wings half-open to reveal his white rump – the 'here's my arse' posture of courtship. His splendour is attained not by moulting into new spring plumage, but by the gradual wearing away of the dull tips of his winter feathers, like a slow unveiling, until he acquires the full livery of some strutting Tudor grandee.

The track curves right towards the river, whose industrial noise is now crashing through the trees. A galvanised steel suspension bridge crosses the river at a narrow point where the waters are forced to roll over themselves in a boiling cataract. The name 'Jäniskoski' (hare rapids) has been cut out of the box girder that supports the suspension cables at this near end; in the same place on the far side is the word 'Njuámmilcooksâs', which I assume is the Inari Sámi name for this place – whether it also means 'hare rapids' or has its own significance I don't know.

I leave the path, drawn across a spongy mat of crowberry, moss and lichen to follow a thin, tremulous call from the treetops 50 yards away. There is a bell-like quality to each minute pulse of its trill, and I suspect it is a waxwing, from memory of the irregular occasions when they descend from the north into Britain, in what birdwatchers call 'waxwing winters'. The forest floor is an obstacle course of vegetated hummocks, and my progress is slowed – and finally halted – by trying to avoid treading on the reindeer lichen. *Cladonia rangiferina* is the staple winter diet of reindeer and alpine hares, but grows only a few millimetres a year; it can take decades to return if trampled. I am unable to proceed, and stand still, listening and taking in details of the scene around me. A pine tree has rooted into the moss on top of a car-sized rock. Now 60 feet tall and a foot in diameter, its four roots grasp the rock like the claws of some giant raptor. Two waxwings shoot through the forest past my head, and there are three more travelling in the same direction at tree-top height. Their flight is arrow-like, their wings sharp-ended. I see their unique colour – dark pink-tinged grey-brown – but they are too fast for me to see any detail of their plumage, other than the diagnostic yellow tips to their tails. They fly far, and I return to the path. The track bears right alongside a pond, where there is a lone female goldeneye. It is raining lightly and the cold surface of the pond registers each dropfall, creating an exaggerated record of precipitation and giving the water the appearance of frosted glass. Under the dull sky and in the dark of the pines, the goldeneye's eponymous insignia is like the first star of the evening. Two mallards fly over, and it is only after they have vanished that I think anything of it, remembering that they may have flown a thousand miles to be here.

24 May, Inari. Lake Inari is clear of ice, weeks ahead of its usual time. For a moment, below my window, there is a ghostly remnant of the ice-time: a white hare.* She sees me, but doesn't move; her instinct for camouflage overrules the reality of the season. She is out of sync. I lean away from the window to reach for my camera, my gaze averted for two seconds. She is gone. She never wanted to be beautiful. She never intended to burn an image onto my mind: the snow-coloured feet and undertail, the glacier-grey flanks, the glacier-white head and ears, tipped crowberry black. Her stare, her absence.

* Mountain hare: *Lepus timidus*.

I return to the forest, which has been quieted by hours of light rain. I take note of the colours and sounds of this early spring – the Sámi season known as *Giđđadálvi*. There is something synaesthetic about the experience: pastel sounds and mute colours. The reddest things in the forest are last autumn's bilberry leaves, an earthenware red, flattened to the ground and cloistered among the evergreen crowberry. Reindeer lichen is blind-eye grey; the erratic boulders, where their moss scales have been sloughed off, are fog-pink. To find any indelicacy of colour, you must look at the shy green birch buds so closely that they fill the view. Step back, and the trees spray their lime-green wastefully into the air.

The waxwing that lands in front of me, in the undergrowth 20 yards away, has brought splinters of yellow, red and white on his wings and has a buttercup-yellow tip to his tail. As he forages for thawed berries, he keeps these regalia hidden, revealing only mammalian shades, mixed from the forest's palette. Only mink traders have names for colours like this: above, he is 'red-glow' – a silky pinkish brown – and 'palomino' below; his rump is 'Aleutian grey'. His black eye-mask and chin and his lax crest appear to be his only extravagances. Then, when he flies up onto a branch, the narrow streak of yellow at the edge of his folded wings, the white tips to his inner wing, in turn tipped with sealing-wax red, the broad yellow tip to his tail, are suddenly incongruous. It is as if he hadn't noticed his adornments, as if they had crawled onto him from the undergrowth to be carried off through the canopy. He flies away, and a tinkling trill recedes with him, a sound as dilute as the landscape.

A Siberian tit arrives into a nearby pine tree. I have never seen one before, and no birdwatcher can feign indifference to a new species: but a controlled fist-clench is the only external sign of my revelry. It is preening 20 feet up, and I find my attention focussed on one detail, the smallest detail of its plumage. It is the black line that runs from the base of its beak and across its eye, and, once past its eye, finishes in a stiletto point. There is perfection in the symmetry of this insignificant mark, a narrow fusil that separates its light-brown cap from its white cheek patch.

Näätämö, Lappi. 69° 40' N. The evening is mild and the sun bright. The border is about 2 miles away, and I decide to take a gentle stroll into Norway before bed. This morning's pine and spruce is replaced by thinning juniper and birch. As I leave the village behind, the three ringing notes of a greenshank reach me from the shallow plaits of the Näätämöjoki river to my left. To my right the land rises by 100 feet or so, and the modest increase in altitude leaves its flat moorland above the tree line. After the road curves to the right I see tall galvanised lampposts and elevated CCTV cameras ahead. Through binoculars I can just read the blue sign that marks the border, and as I look, a bird flies up from the left and perches on an overhead cable directly over the sign. It is too far for me to work out what it is and I continue, stopping occasionally to check if the view has improved. Then it leaves its perch, dropping off the wire in a downward diagonal glide like a sparrowhawk, across the road to disappear into the juniper trees. My walk turns into a march as I fix the spot with my gaze. I feel sure that I have lost the bird, but I am not ready to give up the chase. Suddenly, it reappears, closer, and flying towards me. It lands on the tip of a juniper tree, in full view and in perfect light. An owl, a hawk-like owl; a northern hawk owl, another first. It lets me examine it for half a minute: upright like a hawk, long-tailed, barred below. Only its large, flat head betrays its true affinity, with its smoky-white face bordered in black, its target-finding yellow eyes watching me watching.

Norway

The other name for Näätämö is Neiden, and on this side of the border there is another village called Neiden whose other name is Näätämö. They are now 9 miles apart but were once one village, until 1852, when the border was closed and the Sámi forbidden from herding reindeer across it. I return to Finland for the night, across the cattle grid that now allows free movement of people, their cars, their goods, but not their reindeer. At the edge of the village, two male redstarts are drawing up their own border, civilly with song.

25 May, Neiden, Finnmark, 69° 42' N. I retrace the two miles to the border, the road rising steadily, as it does for another mile on the Norwegian side. When the road curves round to the right, tucking into the side of the fell, the view to the left opens out and 200 feet below is a flat, boggy plain. In the morning sun, I see the burnished-brass glow and casual wingbeats of a short-eared owl quartering the mire. With another 6 miles to the bus, I am grateful to have a reason to stop and remove my backpack and the smaller pack slung across my chest. The owl perches among the birches that line the far edge of the bog, and my attention is taken by something closer at hand, a maniacal chatter and a few seconds' glimpse of a tiny airborne dervish, abandoned to the fervour of a dance-flight that throws him wildly about in the air. He lands in a bush about 20 yards away, tail cocked. A bluethroat, *Luscinia svecica*. *Luscinia* – nightingale; *svecica*–Swedish. The Jenny Lind of birds, named in 1758 by Sweden's best-known scientist, Carl Linnaeus. I watch him for a while. He signals his intent to begin his display with some low-pitched knife-grinder calls. Then he launches into an erratic rise to 40 feet. He sings a medley of four different song fragments, made up of whistles, then rattles, then mimicry and finally 'peeps'. All delivered in a vaguely spiral descent, his body rocking slowly from side to side as he deliberately mis-controls the flow of air around him.

Between here and Neiden, I pass through at least eight bluethroat territories. Their styles vary: one bird sings from a concealed perch a slow version of the usual song pattern. Only when he finishes his song does he launch his aerial dance. Another collapses theatrically off his perch, flashing his coppery tail patches as he tumbles into the dwarf willow scrub, only slightly more slowly than under the pull of gravity. Suddenly, I remember the words of a real dervish, the thirteenth-century poet Rumi:

> Birds make great sky-circles
> of their freedom.
> How do they learn it?
> They fall, and falling,
> they are given wings.

The air is part of the mountain, as Nan Shepherd said of the Cairngorms, and of mountains in general.[1] In a place where only a slick of vegetation coats the rock, the air is a vital third dimension. For a rough-legged buzzard facing into the slender wind, it is a perch from which to watch for voles and the scent marks they leave, which are ultraviolet in the colour spectrum visible to the bird. I watch it hover. Its wings are in a fixed dihedral; its alulae, the feathered wing slats that force air flow over the wing at slow speed, are fully open; its distinctive white, black-banded tail is fanned and twisting, ruddering the air like a kite's.

For the bluethroat and the meadow pipit the air is an arena, compensation for a paucity of trees from which to sing.

For the golden plover, the wood sandpiper and the spotted redshank, it is a void that demands to be filled by their *lalings*, distinct yet remote, like peals from three distant belfries. Part cry and part song, *laling* was still the best way for humans to communicate, too, until 1970, when this road was built. From the 1950s, Bombardier snowmobiles enabled messages to be passed between fells and valleys in winter, but in summer there was no substitute for the Arctic version of the Alpine yodel. People identified themselves using vocal signatures, switching between deep chest-song and falsetto to create a multiphonic bow-wave that penetrated great distances. It is the sound I have been hearing since La Serena, the cry that reached in the moonlight across the Ouse Washes, that made echoes in the spruce bogs of Sweden, and that I heard just now, from a crane somewhere in the mire below, too far away to see.

The bus leaves Neiden at 12.45, and almost immediately bears north-west, gaining height and looking across the great mires to the south, the tundra. The birches are reduced to scrub, where they occur at all, and on the rocky slopes to the right and in the distance ahead, snow, though patchy, is extensive, covering about half the high ground. After a quarter of an hour, between rock outcrops, I glimpse the Bugøyfjord below, and see that the tide is out. A pinnacle rises 500 feet sheer from the sea, and two rough-legged buzzards are in the air at its tip; it appears they are displaying, but as the bus winds on, dropping to the shore, I see them for only a few seconds. There is a Finnish-speaking settlement at the coast, but no one requests a stop, and there is no one waiting to get on; the road climbs again. Where the drop of the land forms an edge against the sky, suddenly a huge shape fills the void – a white-tailed eagle mounting the updraft a few yards away as we pass. This southern fringe of the Barents Sea, a shatter of fjords within fjords, is a pattern of intrusions of sea into mountains, with no coastal plain except where a small estuary has created one. In such places there may be a house or two, and the road drops to serve them before rising back into the limestone. On the elevated stretches there are odd tarns and pools, some with a drake wigeon or two, their ducks presumably by now on eggs.

Another white-tailed eagle, this one on a rock, looking like a rock. There is a thirty-minute stop at Varangerbotn, at the head of the Varangerfjord, and there is a birdwatching hide there, facing along the fjord. There is another white-tailed eagle and four cranes on a pebble skear exposed by the fallen tide. I say the word out loud, involuntarily, recognising a friend: I lived for several years close to Morecambe Bay, and somewhere on this coast is where the Old Norse *sker* set sail in the mouths of Vikings. We head east, and from now I have the sea in view all the time, the journey punctuated by eagles: one at Gornitak, another between there and Abelsborg; one at Abelsborg itself and another at Bergeby; an eighth at Klubbvika, a ninth at Vestre Jakobselv.

Vestre Jakobselv, Finnmark. 70° 07' N. I read somewhere that along the coast from here, north of Vadsø, the most northerly trees are to be found. Beyond that, the fells of outer Varangerfjord, the *fjell* country, reach sea level. As the bus heads north along the coast, the fjord is on my right, and to the left, behind the coastal village of Vestre Jakobselv, the land rises to 200 feet. Against the flat line of the hill, the downy birch trees form a horizontal strip whose upper edge, at this distance, seems to have been precision-razored, with tree-bare moorland above. There is a campsite in the village, and I leave the bus here, wondering what the last woodland in the Arctic might yield tomorrow.

26 May, Vestre Jakobselv. I have descended 390 feet since Inari and 260 feet since Näätämö, and there are wild flowers again. By the shore, two tiny flowers attract my attention for their modest beauty and their unfamiliarity. They are Arctic specialists. The coastal or strict primrose has pale-purple campion-like flowers with a yellow eye, on a stiff erect stem about 4 inches high. It is adapted to extreme conditions and short summers, being less reliant on insects than most of its family. Its stamens and pistils are of equal length so that it can easily self-pollinate. The Arctic yellow violet is as delicate as any of its kind, yet survives because of, not despite, the abundant snow cover: late-melting snow protects the frost-sensitive plant in uncertain weather during spring and early summer. It likes damp, sheltered, shady places less exposed to the wilder forces of nature. But it takes three years for its first flowers to appear, three growing seasons in continuous daylight. Like other violets (so-called, this species being as yellow as a buttercup), its seeds have an elaiosome – an oily cap that attracts ants to drag them underground. The protein-rich cap is eaten and the seed discarded amid the nutritious detritus of ant society. But in the Arctic there are few ants, so the seeds are also adapted to survive the journey through a reindeer's digestive tract. The plant's main adaptation is to produce flowers of two kinds: a beautiful pansy-like flower to attract pollinators in a good insect year, and others that don't open but self-fertilise.

From Vestre Jakobselv, I look up at the *fjell* where for a few days the progress of spring is measurable in the intensity of green that spreads uphill through the downy birches' bandwidth in daily increments. I find a track beside a football pitch and follow it out of the village, uphill and into the trees. They all have multiple trunks, as if coppiced many years ago; perhaps they were, for firewood. Or maybe the first years of each tree's life is a succession of lost battles with browsing reindeer and mountain hares, until one year they gain the upper hand. Perhaps it takes the odd year of early snow-melt and a glut in forage to allow each sapling an extra week to begin its slow maturing. They are short, none more than 10 feet high, and thin, with trunks all less than 4 or 5 inches in diameter. They seem senescent. Most have a few dead trunks among the living, some reduced to brittle spurs of rotting wood. Some whole trees are lacking leaves and show no sign of life; all have dark, bronze-coloured lichen covering the upper two-thirds of their height. In the last few square miles of land with any hospitality for trees, they occupy the outer fringe of their liveable world.

The ground is modestly cloaked in a low, leaf-litter-clinging cover of cloudberry leaves. The first cloudberry flowers, cloud-white, reach singly above the green. The berry itself I know only from pictures and the sting it gives to Sámi cuisine. They are like blackberries in form, and like nothing in colour: apricot heavily diluted with cream, perhaps. Yet another Arctic shade that, if it were to have a name it would be that of the eponymous fruit. Over rocks that protrude above the thin soil there is a clubmoss that grows in snake-like dreadlocks up to a foot long, trailing, hanging over edges, here and there erecting a lime-green inch of new fruiting growth. *Lycopodiella inundata* has the look of the Carboniferous about it. From shaded hollows and ice-cracked fissures, gracile ferns reach delicately for the light. Five inches tall, their stems end in three small fronds separated by equal angles – a trine. Their leaves are curled inwards like the unopened wings of new insects. Having seen one, I can suddenly see thousands, none fully unclenched: today must be their pre-programmed day of emergence. A woodcock squirts up from my tread, and fires itself, quills whistling, between the trunks. A willow ptarmigan, the Arctic version of the red grouse, is still mostly in the white winter plumage that the British grouse never acquires; it therefore lacks the woodcock's camouflage, and rises 30 yards ahead of me.

The wood ends abruptly at a wall of rock – a low, south-facing cliff about 10 feet high that separates the slope from the *fjell* top. I scan the route ahead of me, looking for a scrambled passage, and notice a rodent-like movement among the leaf litter and lichen on a platform of rock. The animal is above my eyeline and obscured from my view as it fossicks towards the back of the short ledge. I know it is there from the scraping and tapping and occasional glimpses of movement. Suddenly it comes to the edge of the rock in full view: a female lesser spotted woodpecker, lifting plates of moss and lichen from the rock to search for insects. It is years since I have seen a lesser-spot in my own country, where it is plummeting towards extinction, its decline thought to be due to low breeding success. They have an interesting breeding strategy, in which females often leave the males to finish rearing the chicks on their own. In continental Europe, males are able to increase their feeding visits to replace food brought in by the female, but RSPB studies show that in England, there is insufficient food density for them to raise the chicks alone, and fewer leave the nest.

I surmount the rock wall that defends the wood from the north wind; a cuckoo is there, calling (a minor third, noted), and four meadow pipits, likely cuckoo hosts, are claiming their territories in song. Bird song is suddenly thinned, but there is a liquid clatter of stones flowing over stone, scree trickling away from the horny battery of reindeer hooves.

27 May, Vardø, Finnmark. 70° 22' N. At 6.00 a.m. I emerge from my tent, which is the size and roughly the shape of a coffin, and which I have shared with my bags and two weeks' worth of clothing. It is 18 degrees Celsius outside, unusually warm, but not unusual, I suppose, for any warmth to have built up over the long sunlit hours and to be felt this early in the day. I pack up my tent and walk down to the coast road to catch the 7.55 bus to Vardø.

In a few miles, the coast road will turn to the north at the wide mouth of the fjord, and until then, I look south out of the right-hand side of the bus, across the fjord to the hills of

Sør-Varanger and Finland, with their dots and dashes of unmelted snow. The sea is flat this morning, but with the sun already high, there is little glare. From time to time I look inland out of the opposite window, hoping to notice where the last trees are. The bus stops in Vadsø, whose street trees are the last; the rest will be tundra.

We cross the Sjåbuselva river east of Vadsø's tiny airport, and I notice a stone circle on the bank; nearby is a white-tailed eagle, itself like a rough-hewn stone block, in a scene that has all the appearance of one unchanged from the Neolithic. But the circle is a World War Two monument. The stream runs off the fell over mineral-rich sedimentary rocks, shales and sandstones and across the shallow marine beach deposits to the sea. As the bus twists across the bridge I just have time to note, at the river mouth, three red-breasted mergansers, a pair of Arctic terns and two goosanders, all specialists in capturing small fish.

I cannot imagine living among white-tailed eagles. I would have to imagine them populating my earliest memories, and every daily scene. Would I notice them? Would I love them? What myths and tales would I have been told about them, and would I believe them? Have they perched on the same favourite rocks and skears generation after generation, century after century? Is it remarked when an eagle fails to return to its usual perch? Are they given names? Are they guests in our world, or are we in theirs? Or neither? Neither. There is a white-tailed eagle in flight as we pass Golnes, another at Krampenes; two on the shore between Skallelv and Komagvær; and two together at Komagvær itself.

The road dips into a tunnel under the sea, and emerges in the island town of Vardø. 'COD IS GREAT' says a 12-foot-high slogan painted on a warehouse across the harbour from my hotel window. It is one of a number of works in a street-art project, and it draws my eye along the line of wharves dotted with kittiwake nests and their guano. As I open my window, a swallow streaks past it, like a welcome companion. I have forgotten to look for swallows in these last few northbound days. I have been hoping to mark the farthest point in my encounters with them, and was sure I had missed the last one.

Hornøya, Finnmark. 70° 23' N; and, at 31° 10'E, the easternmost point in Norway, and of my journeys. St Petersburg, Istanbul and Kiev are all west from here.

From the quayside a few paces from the hotel, a 30-foot Hemnes harbour boat, the *Hornøya*, takes me, Alex Langenegger and Danielle Zaugg across the short sound that separates Vardø's island, Vardøya, from two others – Reinøya and Hornøya, which are inhabited only by seabirds and a relay of scientists. Alex and Danielle are in their sixties and have driven here in their camper van from their home in the Jura Mountains of Switzerland. Alex is a nature photographer and raptor enthusiast, and is compiling material for a book: 'Au Pays des Pygargues' (In the Land of the Sea Eagles). We are the only customers for the three o'clock sailing. Halfway across the sound, we receive our instructions in English, which Alex asks me to repeat in French. We must stick to the footpath. We may see people off the path, but they are authorised scientists. We must be off the island no later than the last sailing at five.

The *Hornøya* slows down for the final 200 yards of the crossing, which is through dense rafts of seabirds, like a slick clinging to the islands' shores. We turn into the narrow sound between the two islands, the larger Reinøya to our left, the 170-foot cliff of Hornøya rising sheer to our right. Guillemots fly-run across the surface as we cut through the flock, some taking to the air, most breasting the water again as soon as they are out of the way. There is a small landing stage at the base of the cliff where half a dozen people, most of them with telescopes and telephoto lenses, are waiting to be taken back to Vardø. A few others have chosen to stay and wait for a later sailing. Viewed from the sea, there seems to be nowhere to go once we land, other than to cling to the base of the cliff, but once ashore, I realise there is a rim of flat land – a raised beach – and a path leading off to the right into the interior of the island.

Hornøya is 99 acres in extent, shaped like a balloon squeezed into two unequal bulges. The larger, northwestern bulge is dominated by an oval hill, whose seaward flank is the cliff under which we have arrived, home to 14,000 guillemots, 10,000 kittiwakes, 15,000 puffins and a few hundred razorbills and black guillemots. From the 213-foot lighthouse-topped summit, the island falls away in an undulating slope to the south-east. These Arctic waters mix with the Gulf Stream and are permanently ice-free, attracting seabirds back early to profit from the rich fishing of the Barents Sea. The productivity of these waters allows room for an additional species not found in the seabird colonies of the British and Irish coasts: Brünnich's guillemot. It is a bird of the high Arctic, and we are at the southern edge of its range here, with about 500 pairs scattered through the colony. As I crane my neck and take in the enormity of the high-rise city above me, I realise I will have to scan each cranny and fissure if I am to spot any Brünnich's among the throng.

I stand at the edge of the raised beach, where the ground is thick with the shiny green leaves and small, white cress-like flowers of scurvy-grass, prospering in guano-enriched soil. The overhead traffic between the sea and the urban terraces of the colony is congested and constant, the noise a tumult. It is this sound that keeps me from my search for the Brünnich's guillemot. It seems orchestrated: along the cliff far to my left, the sound changes. The north end of the colony has become louder, audible as a distant ferment against the din of the nearby quarter. The commotion spreads like a wave towards me, almost visible, like a ripple of air along a noise curtain shrouding the rock wall, or a long, held chord spreading through an orchestra. As the wave gets closer, I see its conductor, a raven, quartering close to the rock; as it passes, all the kittiwakes in the vicinity yell their fear, and the salivary gargle of the guillemots rises in pitch and volume.

I take the path south along the cliff-foot towards the right-hand end, listening for new sound waves that always seem to start from the far end and get closer. Occasionally the wave stops, and a new sound takes over, the aleatoric sound of blind panic indicating that the raven has landed. He has spotted an unguarded kittiwake nest, or bullied its owner to move aside. I watch as a pair work the cliffs together, landing directly opposite me 30 yards away. There are two nests inches apart. A raven lands, prompting one nest-owner to jab its bill to the left, stepping off the nest. The second raven reaches in, and in an instant has a jade-grey kittiwake egg between its mandibles.

Where the path winds up a rocky natural staircase, the green, mineral eyes of nest-fast shags follow an assortment of boots passing inches away. Before I leave the colony behind, I scan the ledges from a more elevated position, finally catching sight of the white streak along the length of its bill that marks the Brünnich's apart from the common guillemot. There is a group of twenty, their backs towards me as they guard their places on a ledge.

In the high Arctic, the Brünnich's is an abundant bird, with about 850,000 pairs in Svalbard 350 miles to the north. There, common guillemots suffered a population crash in 1986 and 1987, a result of a collapse in the capelin stock, whereas the Brünnich's guillemot increased by 20 per cent at the same time. The common guillemot seems more vulnerable to the fortunes of particular food species, while the Brünnich's is able to exploit a wider range of prey.

Brünnich's spend most of the year living close to the edge of the sea ice, and this appears to be the main habitat difference between the two species. Studies suggest the Brünnich's adaptability may make it less vulnerable to the dramatic shrinkage in sea ice than might be expected. This year the ice loss seems about to break yet another record; as the seas warm, the two guillemot species may find themselves competing for diminishing resources.

As I walk up the slope on the far side of the hill, I bump into Alex and Danielle and compare notes. I have never had cause to learn the French names for Arctic species, making for a bizarre conversation based on descriptions rather than names. As we are speaking I become aware of the sound of another predator reaction, and I am slow to look behind me. When I do, I catch a brief glimpse of a large falcon carrying something the size of a puffin. '*Faucon*!' I shout as I whip my binoculars to my eyes; I get a three-second view of a powerful bird whose tail and wings are narrowly and densely barred slate-grey and whitish. Alex missed it, and Danielle wasn't looking.

The gyrfalcon's name comes from the French *gerfaucon*, and this time I was able to tell Alex what he had missed. It would have been a new species for him, as it was for me. In mediaeval Latin, it is *gyrofalco*, probably a reference to the species' buzzard-like circling as it searches for prey, very different from the power-dive of the peregrine. It is found across the Arctic; in 2011 Kurt Burnham and Ian Newton revealed that birds satellite-tagged in Greenland spend considerable periods of the winter on sea ice, hundreds of miles from land. The gyrfalcon may be another species whose fortunes are set to change.

28 May, Vardø. I lower myself into the basal remains of a World War Two observation post sunk into the tundra, concrete rendered a Quaker grey by millions of microscopic black lichen colonies. It is a harbour for refugees from past and future gales: a fishing net float, a willow shrub. Its microclimate gives a marginal advantage to the cloudberries, which grow 3 or 4 inches taller here, and moss campion, which finds sanctuary in the cloudberries' lee. The campion is also called the compass plant, for its dome-shaped form which sprouts purple-pink flowers like miniature carnations on its south-facing side in early spring. As the season progresses, and light is available continually on all sides, the dome turns from green to pink, as flowers spread over its whole surface.

The bunker's walls are the perfect height for resting my elbows as I scan the Bussesundet, the 1,800-yard-wide sound that separates Vardø from the Varanger mainland, which is still dotted with patches of snow. The land curves down to the shore, where a raven is tackling some prey item. An egg, I think at first, then I remember the shards of sea urchin shells I have been noticing scattered along the spine of the island wherever there is a rock or hard track on which to crack them open. Kittiwakes are shuttling overhead with nesting material collected from the same west shore, over the island, over the tiny islet of Svinøya (swine island) and east over the 1,500-yard sound of Reinøyasundet, to their colonies on Reinøya and Hornøya.

Suddenly I hear a muffled clamour from the west that wells like the sound from a distant football stadium when the home team scores. In the Bussesundet is a small island: Tjuvholmen (thief islet). I watch from the bunker as an adult white-tailed eagle flies over it, prompting a chorus of vocal abuse from shags and herring gulls. The eagle shows no interest and flies north – my right – over the far tip of the islet; there I notice another, perched on the rock below, unmoved by the bird overhead.

I continue towards the island's northernmost tip at Skagodden, a name meaning 'rocky point'. The tundra vegetation is tight to the bedrock, and half a dozen golden plovers in their breeding plumage of yellow-spangled above, white-bordered black below, are alert, as if aware of their prominence in the landscape. Familiar birds (a dunlin and two meadow pipits holding territory) join unfamiliar ones (a bluethroat and three red-throated pipits); they are all among the last birds to have completed their migrations – if indeed they have. Skagodden is a headland cleft in two, like a stone-pillared gateway 50 feet high and 50 feet wide; looking between the pillars is like looking out from the open doors of a sea cathedral. The rocks are clad in one of the Arctic's few long-stemmed plants, the succulent roseroot *Rhodiola rosea*. It is 10 inches tall and fleshy, and has several stems growing from short, thick rootstocks that lodge in tight cracks in the rock – a relative of stonecrops. Two ravens are preaching into the chasm between the pillars from their stone pulpit, and the cliffs respond with a stony resonance.

I return south, towards the town, along the east shore. Throughout the island there are wooden frames – fish drying racks – like series of interconnected goalposts. They are about 10 feet tall and some are 60 or 70 yards long and half as wide, made from grey birch staves preserved by the cold, salt air. There is no wood of the required length, diameter and straightness growing within 100 miles of here, and I guess the wood is continually re-used whenever repairs are needed. At the foot of one of the racks, woolly willow *Salix lanata* grows; it is a low, many-branched shrub, 2 to 3 feet high and twice as broad with pale grey-green leaves covered in silvery-grey wool. This year's new twigs have begun to sprout, growing hairy at first for protection against the chill; thick, 2-inch-long catkins have attracted a few bumble bees. From this sparse cover come intermittent explosions of song from two redwings, two willow warblers and two bluethroats, like sudden bursts of flavour, sweet and acid.

A raven flies across the Reinøyasundet to Reinøya, and draws my eye to yet another rock-like white-tailed eagle at the shore; then another wave of crowd protest washes back as the raven approaches the island.

In April 1848 Lorenz Brodtkorb and some companions were rowing from Vardø to Reinøya. In the Reinøyasundet between the two islands, they came upon four swimming birds the like of which they had never seen before. Brodtkorb shot and killed one, which he later described as 'the size of a brent goose but like an auk in shape'. He noted that the wings were so small that the birds could not fly off when his boat drew near. Brodtkorb dumped the corpse on the shore, intending to retrieve it later, but when he returned the following day a storm sea had washed the bird away. He set out in his boat to try to find the three surviving birds, without success. There can be little doubt about the bird's identity, a full four years after the date usually ascribed to the killing of the last great auk, at an offshore rock in Iceland.

Båtsfjord, Finnmark. 70° 38' N. At Vardø harbour I board the MS *Nordkapp* of the Hurtigruten line, the Norwegian coastal express, for the 4.45 sailing. The 440-foot, 11,000-ton Polarlys-class ship is on route between Kirkenes on the Russian border and Bergen, a five-day voyage. She lies alongside the quay for half an hour, and once under way, is escorted into open water by a white-tailed eagle. For the three-hour voyage west to Båtsfjord, we are accompanied by a relay of seabirds: thousands of guillemots, puffins and razorbills from Hornøya and a chain of smaller colonies to the west; two gannets from the world's most northerly colony at Syltefjordstauran; and six fulmars – three white and three blue. I have never seen a blue fulmar before, the smoky, midnight-ice-coloured form that makes up a proportion of the Arctic population.

Another pair of white-tailed eagles are there to escort the *Nordkapp* as she sails down the 8-mile-long fjord and docks at Båtsfjord, the largest fishing village on this coast, with 2,100 inhabitants. The deep water of the fjord affords anchorage to a ship like the *Nordkapp,* while the smaller inner harbour is the base for a fleet of fishing boats. We are on time at 7.45 p.m., giving me four and a half hours to scramble up the nearest hill and back before another ship arrives, serving the same route in reverse. At 700 feet, the pyramidal Melkarn overlords the harbour in the west, and shades it from the evening sun. From the outer quay I walk round the inner harbour and into the village looking for any track leading up and to the west. Unlike Vardø, Båtsfjord has trees in the shelter of the hills and houses, and as I walk up the lowest slope of Melkarn, up Kirkegårdsvegen and past the town's 1910 red-brick church, a male redstart is singing a rough-voiced serenade.

There is a path leading uphill off Kirkegårdsvegen at the upper edge of the town, like a narrow scar cut into the skin of wind-razed shrubbery. There are trees, of a kind – long, twig-thin procumbent dwarf birch that scramble over the shale and through cushions of crowberry. Double pink bells of bog bilberry and russet-tinged chickweed wintergreen leaves weave colour through the nap, and scattered flashes of purple-pink mark early outbursts of alpine azalea. The incline steepens, and in places segments of the path are connected by short scrambles over rock outcrops, out of whose more sheltered alcoves spring sprays of alpine lady's mantle leaves. At one such place, where the rise of the ground and the bend of my body brings flowers to within inches of my face, I fancy I catch a waft of fragile perfume, gone as soon as it is half sensed. I see something bright at the corner of my field of view, a cluster of six mountain avens flowers shining white against the shade. When I reach a flat few square yards of rock, I stand and notice several patches. They all have eight petals, giving them an unusually circular star shape. Their specific name *octopetala* refers to this oddity, while their generic name, *Dryas*, is both borrowed from and loaned to bygone eras. Dryads were tree nymphs in Classical mythology, divine spirits who animate nature, bound to their immediate surroundings during their supernaturally long lives, and depicted as beautiful, young nubile maidens. These associations must somehow have suggested a name to Linnaeus. In turn, three periods – or stadials – of especially cold climate towards the end of the last ice age have been named the Younger, Older and Oldest Dryas stadials. In the taiga zone, forest retreated and was replaced by tundra, returning and receding with each oscillation. Each taiga phase is identified by the high levels of mountain avens pollen found in soil core layers from those times.

There has been an unfulfilled threat of rain in the sky for most of the evening, but as I near the summit of Melkarn, it takes on a peculiar luminosity. From this deeply shaded side of the mountain I look over Båtsfjord far below, and across to the cliffs and hills on the far side of the fjord. The sky is the same rain-engorged purple-grey, but illuminated by an unseen light from behind me. There is a patch of snow a few yards square at my feet, and I find a rock pew to sit on to catch my breath and take in the newly polished scene. I catch sight of a movement to my left, and turn my head in time to see a bird fly out from under the far end of the snowdrift and land on a rock 30 yards away. It is a ring ouzel, the blackbird of the mountains. It turns to face me, its white breastplate like a map of the crescent-shaped patch of snow it just rose from. Then it is gone, downhill, hugging the slope. I don't bother to check the time, aware that should my ship appear to the left at the mouth of the fjord, I would have time to get back down the mountain and through the town to the quayside to catch it before it sails at a quarter past midnight. There are a few more feet to climb, and when

I lift my head over the rock parapet at the summit, the sun's dazzle is almost physical, like a slap in the face, and I can only look downwards at my feet until my eyes adjust. It is bare rock on that side, and the sun's glare comes both direct and reflected off the plateau.

I am back at the quayside; the MS *Finnmarken* is delayed, and there is no one there to tell me by how much, or indeed if it is coming at all. I look at the signs on the wall of a hut, wondering if there might be a phone number to call. I entertain thoughts of building a bivouac out of the stacks of yellow 100-pound fish boxes that line a warehouse wall at the back of the quay, or of walking up a different mountain, one that faces into the midnight sun this time. Eventually, I hear music approaching rapidly; accordion music, Norwegian folk music. A small, yellow Jungheinrich fork-lift speeds through the gates to the quay, turns sharply left and parks by the hut. The music continues from its cab while the driver sits, waiting. When he finally gets down from the cab, he is surprised to see me.

'*Venter du på båten*?'

'Sorry, do you speak English?'

'A little. Are you waiting for the boat?'

'Yes.'

'It's coming.'

He asks me where I have been staying in Båtsfjord, and I explain that I have only been here a few hours, enough time for a leisurely walk up the mountain, before returning to Vardø.

'Melkarn. I think it gets its name from the first people who came to Båtsfjord, who kept their animals there. You know, from "milk". There are a lot of machines up in those other hills, did you see?' I did; I came down by a different route, past a crane and two diggers, along a track newly cut into the hillside. 'That is for windmills. There will be seventeen of them, and a total of forty-seven in the district.'

'Is that a good thing?'

'*Nei.*'

'It will change the way this place looks.'

'Yes, and they are destroying all the vegetation with their machines.'

I discover that my new friend is called Rouald, and that he has lived in Båtsfjord for forty years.

'I've worked here all that time. Maybe another two years. I'm sixty-five; you can stop when you're sixty-seven. Anyway – I must …' The ship docks, and Rouald catches a rope thrown from somewhere.

29 May, at sea off Molvik. 70° 43' N. On the floor of the Barents Sea lies a small piece of Africa.

It is a stone, flat and almost square, about two inches by two; I dropped it from the fourth deck of the MS *Finnmarken* as she drew level with the Makkaur lighthouse at one o'clock this morning. I had been holding it for some time, turning it over and over in my jacket pocket, waiting. I held it at an angle to the white light of the night sky, tilting it to see the ripples on its silky phyllite surface, like an unbroken sea. It had the same satin sheen that I saw when I first held it, as I lifted it from the dry *arroyo* on the slope of Monte del Renegado.

As my journeys reached their most northerly point, I moved to the port side, 65 feet further north still, and released the stone. I watched as it hit the sea, and watched the small, brief scintilla of white spume it formed before it was swallowed by the wash from the ship's bow.

Looking north from the fourth deck, I see the Arctic Ocean stretching away to the pole. What my eyes see is limited by the curvature of the earth, but in my mind the view extends past the horizon to where the sea ice still burns with a faltering glimmer. Unseen, but more than imaginary, it is something sensed. There is only Svalbard out there, most of it lying west of my line of sight, but my gaze runs due north, across the 5 square miles of distant Abeløya, past its neighbours in Kong Karls Land, an archipelago reserved strictly for polar bears. How will this year's record-low ice cover affect them? Will they adapt and learn to live in the new Arctic of warmer seas?

Of the two billion birds that left Africa, how many are tending their young in Spain and France, guarding their eggs in Britain and Sweden, singing their land-claims in Finland and Norway? A few thousand are completing their journeys only now. Ringed plovers, sanderlings, dunlins and turnstones have been tracing the western seaboards of Africa and Europe, burning their fuel between their stations at the great estuaries of the Banc d'Arguin, the Tagus, the Waddenzee, the Wash, Morecambe Bay, the Solway. Only birds with extraordinary life stories live north of here. One, the rock ptarmigan, lives through three months of darkness. There are ducks and auks and fulmars whose lives are lived on the sea, outside the few weeks of the Svalbard summer. Gulls that move with the ebb and stretch of the sea ice. The Arctic tern arrives from the Antarctic, and sees more light than any other animal in the two summers of its year.

Two songbirds nest in Svalbard. To snow buntings, in a land of which 60 per cent is covered by glaciers and 30 per cent bare rock, territoriality is cardinal. So much so that the males arrive weeks before the females, braving April temperatures of minus 30 degrees Celsius, compelled by competitive forces to occupy any of the few hospitable areas. For winter they migrate only as far as the northern temperate parts of Europe, and see few days of double-digit temperatures their entire lives.

Only one land bird brings the sweat of Africa to meet the ice of the high Arctic. The northern wheatear has one of the largest ranges of any songbird in the world, with breeding grounds in the eastern Canadian Arctic, Greenland, Eurasia and into Alaska. They all winter in Africa, following the longest routes undertaken by any small land bird. Even Svalbard is not the farthest point from their winter quarters: birds fitted with tiny data loggers in Alaska have been found to migrate more than 18,000 miles a year using routes they inherited from the Pleistocene, before their breeding range expanded to include the Western Hemisphere. A few weeks from now, young wheatears that have never seen darkness will carry the light of the Arctic across the limestone peaks of the Atlas and over the burning sands of the Sahara to seek sanctuary in the Sahel.

I wonder how long it will take for my stone to settle; and whether the water lies still down there. Perhaps it is destined to wash ashore, its geometry transformed from a flat square to a sea-rounded disc. A secret souvenir of a mountain from whose peak someone once looked south, across another continent, waiting for spring to fly north.

Missed Connections

31 May, Jiepmaluokta, Sápmi (Alta, Finnmark, Norway). I think I am becoming obsessed with maps and lines, trajectories and convergences, and connections.

And geology. This is something I have little knowledge of – my earth-love is skin-deep. My gaze locks onto the folds and curves of its physique; I am easily mesmerised by the glittering shawls and cloaks of its forests and meadows, and glimpsed hints of naked savagery. But here, now, looking across the raised beaches of Jiepmaluokta, down to present-day sea level and over the fjord to the mountain Seilandstuva, there is a part of me that wants to recalibrate time so that I can animate rock and study its behaviour. I would see the isostatic rise and fall of its breathing and hear its shuddering tectonic gasps. Feel the intense heat of its eruptive heaving. The laying bare of its life story: the ephemeral vulnerabilities of its youth, its ageless structural beauty. The igneous, born of fire and the slow cooling of passion. The sedimentary, laid softly flat, as on fine silk, silt-fine skin upon skin, slowly pressed and stilled. And the metamorphic, arching and clattering, squirming and fidgeting.

There is a great, hump-backed rock lying stranded in front of me. Its back was scarred over thousands of years by glacial scraping, then soothed at the water's edge when the ice retreated, reducing its wounds to fine lines like the pleats in a whale's maw. It was lifted from the sea by isostatic rebound, the rising of the land freed from the pressure of a mile-thick ice-cap. It now lies 100 feet above the shore, its body barnacled by lichen, half-sunk into the soil that still grows imperceptibly deeper each year under a shagpile of crowberry, cloudberry and bog bilberry. The path I am on runs past this rock and winds down to the left. Each step down the path to the sea represents a decade or two along a timeline of changes that began about 14,000 years ago, when the ice started to withdraw from this coast. Two or three thousand years later, the fjords were ice-free. Another millennium or two on, about 10,000 years ago, the perennial ice had gone from all of northern Norway.

Fourteen feet nearer to sea level, I arrive at another group of massive, sea-smoothed rocks that emerged from the waters about a thousand years after the stranding of the hump-back, by which time people had arrived at these shores. They discovered a place imbued with a magic or a utility we cannot perceive today. They gathered here from distant settlements, and the shoreline rocks became their message boards. They chiselled images into the fine-grained, grey-green sandstone using quartzite nibs: reindeer, moose, whales, bears and people. I look across what is now wet heathland, over dense, low-growing shrubs and scattered trees. Clumps of Labrador tea, a native Arctic *Rhododendron*, have newly opened clusters of white star flowers shining against their dark-green leaves. Beside me is a great pale-grey rock, the earliest of the message boards, covered in red images. It has been exposed from beneath a carpet of moss by the excavations of the past forty years, and the petroglyphs carefully highlighted with a harmless red pigment for easier viewing. Archaeologists believe that the images were created at the edge of the sea, 85 feet below their present-day position. Then, the rocks would have appeared red, like the ones on today's shoreline, where the salt spray reacts with the rock surface to produce a natural red coating. The glyphs would have reflected almost white, so I am seeing them as negatives, in a way. The exposed rock measures 35 feet by 16. It is not flat, but has a relief of its own, like a model of the landscape surrounding it. The images are placed on its miniature fells and in its miniature valleys like figures in a landscape. Any projection of the scene onto a flat surface would show the figures at odd angles to each other, but in three dimensions they are always oriented the right way up. A herd of wild reindeer is being corralled: there are thirty inside the fence, where a man wields a stick to keep them under control. Six more are at the entrance to the palisade, walking in, and thirty others are browsing freely, scattered throughout the landscape. Four bears – a male, a female and two cubs – have walked across the scene from their den, leaving their footprints in the snow. On the far side of a miniature mountain, seven moose roam. In the distance, eight people form a line, and I imagine them hollering to drive the reindeer towards the corral; nearby, others wield clubs and sticks to keep the bears from the deer. Exactly when these first images were made is a matter of debate: 6,000 years ago based on the rate of isostatic rebound, but recent thinking suggests they may be 1,000 years older. There were times when the ice-melt caused the sea level to rise so much that it cancelled out the land rise, and the coastline may have changed little for more than 1,000 years. At other times, relative land rise was so rapid it would be noticed during a person's lifetime.

Is it a map, this rock; one that shows the best place to raise a palisade to corral the reindeer, the location of the bears' den, the forest where moose may be hunted? I am naturally drawn to any utilitarian interpretation, but the accepted view is that this is a document of the prevailing myths of the time. The bear tracks lead out from the den in three directions: uphill, downhill and horizontally. The upward tracks suggest a link to a higher world; the downward-leading tracks indicate a connection to the underworld through a natural water-filled crevice. I find myself resisting archaeologists' habit of interpreting the unknowable through magic, myth and ritual. On the other hand, here at Alta there are three other depictions of bears roaming up, down and horizontally. And bear cults persist to this day among some Arctic peoples, such as the Nivkh in north-east Russia, and at least until modern, Christian times, the Sámi. These are the earliest known images linked (assuming they are) with such a cult. They are also the earliest known images documenting the reindeer herding that is the defining tradition of the modern Sámi.

There is a museum up the hill, where I collected a guidebook before descending this path through heathland and time. In it, my line of hollering beaters is described as a procession or a dance, a rite or a game. 'Perhaps,' suggests the guidebook, 'the fence symbolises a frame surrounding a cosmic world, the human figure inside it is a spirit or god, and the animals likewise.' I continue down the path to the next rock panel, similar in size to the first and belonging to the same early phase of carving. At the far top right corner is a small boat with two figures on board; a very long line extends down to where a huge fish, undoubtedly a halibut, is caught. Next to it, apparently on the sea floor, is a bear. As the guidebook points out, these are 'figures that do not usually belong together in nature. They may belong together in a mythological story.' I wonder if they belong together in the way paintings in an art gallery do – it's where they happen to be, thanks to circumstance, history and convenience. But however hard I try to rationalise them in my own terms, and to reject the educated fictions offered by the sum of scholarly wisdom, I find myself bound by their spell. Magic or not in their day, they speak some special language to me now. This they have in common with the greatest art.

To walk this path is to travel in time, downhill to the present. The next rock, lying about 60 feet above the shore, emerged from the sea between 4,200 and 5,300 years ago. There are ghosts on this rock, hieroglyphs of gone birds, with messages to the future. It is a small rock, with a handful of reindeer, and three high-sided boats with moose-heads at the prow. They are a different style of boat to the ones the halibut fishers used a thousand years ago (I should say, a thousand years *before*), 50 steps back up the hill of time.

There is a bird, a cormorant probably, standing with its wings open, drying them as cormorants do. And there are great auks, four of them.

I arrived at Jiepmaluokta, the Sámi name for this place, which means Bay of Seals, at 6.30 this morning, an hour after taking off from Vadsø in a Bombardier Dash-8 turboprop. My northward journeying ended two days ago off Molvik, and this was supposed to be a day of pure private tourism, too good a chance to miss before flying back to the UK for the last time.

I was not expecting to come face to face with great auks. I have read about this place, and I have read about great auks. None of the descriptions of the rock carvings mention the bird, and none of the accounts of the bird mention the place. The guidebook has a photograph of the auk-glyphs, describing them as 'probably geese'. There are geese here, too, but the auks have stouter bills, shorter necks, upright postures; one is flapping its paddle-wings, quite de-adapted for flight, and has a man's hand round its neck. It is a huge bird caught by a small Stone Age man. It is either a very bad drawing of a goose or a pretty good one of the extinct great auk.

I stop trying to understand the purpose and meaning of the art on the rocks. I need to untangle the thoughts. There is a black guillemot in the bay, the most beautiful of the extant auks. I see it, 60 feet below my position beside the auk-rock. The black of its plumage is a soft, velvet black with a suggestion of deep red-brown, the colour of black hair in summer. It has a broad white wing band; a red spark flashes from its legs when it dives. Did its ancestors once dive with *Pinguinus impennis*? Did *Pinguinus impennis* live here, the very bird brought home by the hunter whose portrait is beside me? I could run my fingertips along the groove of the great auk's neck, but resist the temptation. Just seeing the thousands of quartzite bites in the image, and imagining a Stone Age finger brushing the chippings away, to nourish the soil at the base of the rock, to create the carpet of dwarf cornel flowers at my feet, is to connect to the ghost-bird.

By the mid-1500s great auks were already almost extirpated from the eastern Atlantic, where their colonies were regularly ravaged for their thick down. As they became rarer, their skins became collectable and their eggs highly prized. In July 1840, Britain's last great auk was captured on the remote islet of Stac an Armin, St Kilda. Three men kept it alive for three days, until a fierce storm arose. Believing the auk to be a witch and the cause of the storm, they beat it to death with a stick. The last known breeding pair were found incubating an egg on the island of Eldey, Iceland on 3 June 1844, and were killed to order at the request of a dealer. One of the captors, Sigurður Ísleifsson, explained how he and his colleague Jón Brandsson crept up on one of the birds. 'I caught it close to the edge – a precipice many fathoms deep. Its wings lay close to the sides – not hanging out. I took him by the neck and he flapped his wings. He made no cry. I strangled him.' Eight years later, the last known great auk was seen briefly on the Grand Banks of Newfoundland.

Pinguinus impennis and modern humans were destined to coexist only for a short period between the ice age and the threshold of the conservation era. Had it hung on another fifty years, it might still exist today, perhaps even restored as a common bird. It was eventually wiped out not by the necessity of easy meat, but by avarice. It is officially the last European bird to suffer worldwide extinction, but it is highly probable that another, the slender-billed curlew, has already followed it to oblivion. It has become a victim of hunting during the age of unnecessary hunting, just as we were developing the miniaturised satellite tracking technology and finding solutions that might have saved it.

24 June, London, United Kingdom. I awake in a different country.

I went to bed at 2.30 this morning, when the early results of the EU referendum suggested Vote Leave might win, but at a point when Remain had nudged in front. At 6.30 am I catch a train to London for a meeting. My laptop remains in my bag as I allow the gentle undulations of South Yorkshire, Nottinghamshire and Lincolnshire to provide a familiar backdrop to my unfamiliar thoughts. A journey of any kind, and a rail journey in particular, invites thoughts about connectivity. This spring I have crossed thirty-four degrees and forty-nine minutes of latitude and, I realised only lately, an almost identical spread of longitude. Most of the distance has been covered by train and bus, watching strips of landscape as it evolves, like a self-narrating visual poem. By May, as I was approaching the Arctic, I began to think of the journeys as verses in that poem, stimulated by talk of *Kalevala*, and earlier of Mabinogion. I have been struck by the power of epic poetry to cement an idea of a land, a people, a tongue, defined by all the things that borders cannot contain, and none of the things for which borders have a purpose.

Perhaps Europe's problem is that for all its creation myths and even its anthem (Schiller's poem set by Beethoven is well chosen), it lacks a national epic. That said, there is real poetry in an orange-covered tome that I have on my bookshelf. Published, fortuitously, in 1989, the year the Berlin Wall fell, *Important Bird Areas in Europe*[1] describes 2,444 sites* with a kind of incantatory resonance, one verse per site. Each takes the common themes of place, nature and circumstance and develops them for the diverse landscapes of Europe, from the Azores to the Urals. Regularly updated,

* The most up-to-date inventory lists 4,887 sites.

it remains the fundamental reference for bird conservation efforts across a continent. Many of the places I have visited this spring are in there; some owe their continued existence to the recognition this confers; some are in deep trouble in spite of it.

Ten years before *Important Bird Areas in Europe*, in the Swiss town of Bern, nineteen nations signed a Convention on the Conservation of European Wildlife and Natural Habitats, recognising that we shared responsibility for the wildlife of our small continent. Nine of them were the countries of the European Economic Community, as it was then, who had begun to recognise that birds have always enjoyed a border-free Europe. Wildlife conservation was one powerful reason why common standards and co-operation between countries needed to extend beyond the purely economic realm. The EEC (now EU) Birds Directive was born, followed in 1992 by the EU Habitats Directive. As more Central and Eastern European nations emerged from the former Soviet Union, the former Czechoslovakia and the former Yugoslavia, so the consensus that had coalesced around the Bern Convention grew to forty-five states, while the EU grew to twenty-eight. The Directives and the Convention have come to symbolise shared space, shared heritage and free movement of the most ancient of Europeans.

In recent years some parts of the business and political communities – the distinction between them is often unclear – have sought to override the Directives. In 2014 the EU's political leaders appointed a European Commission President, Jean-Claude Juncker, who had a personal mission to make the Directives 'fit for purpose'. Many of us saw this as thinly veiled code aimed at putting the interests of developers, investors, farming lobbyists and the politicians who are in thrall to them above the interests of the wider citizenry. The Commission set up a team to conduct a review, believing it would amass the evidence for reform that they needed. It did the opposite. Earlier this year the so-called Fitness Check confirmed that 'the balance of the evidence shows that the Directives are fit for purpose, and clearly demonstrate EU added value'. A ruffled Commission sat on the report for several months before it was released under Freedom of Information legislation. Meanwhile, 520,325 EU citizens responded to the Commission's public consultation by opposing any weakening of the Directives.* Ironically, there are two reasons to question whether the Birds and Habitats Directives – and for non-EU countries the Bern Convention – are indeed fit for purpose.

Firstly, they were supposed to set minimum standards that would apply across Europe, explicitly inviting countries to go beyond the minimum. In practice, no country meets those standards, and most, including the UK, don't even come close. The unintended consequence of having strong, well-drafted legislation is that, far from setting a minimum, it has at best defined an aspiration. Nor is there any attempt to pretend otherwise. The official target for protected areas set by the governments of England, Scotland, Wales and Northern Ireland is for 95 per cent of Sites of Special Scientific Interest (SSSI)† to be in 'favourable' or 'unfavourable recovering' condition by 2020 – comprising 50 per cent favourable, 45 per cent unfavourable recovering. The original date set for achieving this goal, 2010, was missed by a mile. In this context, 'recovering' is a sham: it means that 'under current management conditions the site is likely to become favourable over the course of time'. The time period is not specified, and management is invariably

* On 7 December 2016 the European Commission finally confirmed it had shelved plans to amend the Directives.

† Known as Areas of Special Scientific Interest in Northern Ireland.

based on short-term agreements under schemes whose future is uncertain. To obfuscate further, the government often uses 'target condition' to cover a combination of favourable and unfavourable (but supposedly recovering); and this has been further spun into 'the very best condition' in at least one government report. Thus it takes just three orders of sexing-up to move from 'failing to meet the minimum standard' to 'as good as it is possible to be'.

At site level, the obfuscation borders on the dishonest. In 2016[2] the Dark Peak SSSI is shown as 97.73 per cent meeting the target condition, but that figure comprises 4.33 per cent favourable, 93.41 per cent (amounting to 71,340 acres) unfavourable, and allegedly recovering. Drill deeper: on some of the largest 'recovering' units, assessments include statements such as: 'Some burning* is located next to gullies, the gullies look to be used as a fire break. This is having a negative effect on the re-vegetation of those gullies. The unit is dissected by many active gullies, some of the larger ones are down to bedrock.'[3] So a site can be 'recovering' when on-going bad management practices are clearly identified by government assessors.

Secondly, as well as promoting areas for special protection – Natura 2000 – the Directives also provide a basis for maintaining healthy wildlife across the continent as a whole, outside protected areas. Some species are, or were, wider countryside species by definition: you cannot realistically create nature reserves for skylarks, linnets or cornflowers. In the years since 1979, wildlife outside protected areas has plummeted, due mainly to EU-funded agricultural intensification.[4] In southern Britain, birds like snipe and lapwing have become associated with protected areas because they are no longer found anywhere else. When I was growing up in Kent, it would never have occurred to me to visit a nature reserve to see lapwings.

* of heather for grouse management.

Still less so during our family visits to the Fens, which, in little more than a generation, have become devoid of the species whose air-filling calls once defined the Fenland experience – a version of which is now to be had only in nature reserves.

Conservation NGOs have been forced to become complicit in the lowering of expectation and ambition. The safety net has replaced the trapeze as the place where most conservation advocacy happens – fighting to save Directives; campaigning for existing laws to be enforced; devising Schemes for pockets of wider countryside. Working around incompetent, uninterested or wilfully destructive governments has become the standard mode of engagement, leaving little room for any fundamental reimagining of our relationship with nature, for reconstruction.

Take hunting. Calls for control take one of two forms – oppose it outright on moral and/or cruelty grounds, or take a morally neutral stance and manage it to minimise its impact on species and habitats. Neither considers the cultural implications in this twenty-first century of giving individuals proprietorial rights over our impoverished common heritage. All the traditions and practices of hunting in Europe date from a time of abundance. In Sweden, bears, which are protected under EU law, are lawfully killed, spuriously citing the Directives' provisions for livestock protection to assuage a minority field sports lobby over a majority of objectors. In Britain, 40 million non-native pheasants are released every year. Millions are killed on roads, nourishing a population of crows and buzzards that are then shot to prevent them depredating next year's stock of poults (the former legally, the latter either illegally or with secretly issued government licences). The surviving pheasants are shot for fun. Releasing any other non-native species into the wild would be illegal, so for convenience pheasants are classed as livestock. Grouse shooting has enjoyed increasing public subsidies despite being one of the main causes of conservation failure in British National Parks, where vegetation is denuded, landscape uglified, species diversity reduced, raptors eliminated and water catchments buggered.[5] In France, making hunting the right of everyone anywhere as a tenet of the revolution has defined it as a cultural heritage. Now, 40,000 French hunters are giving up each year because they have nothing left to hunt outside special, expensive areas, many of which are supposedly protected under EU law for other species. The case for re-evaluating the place of hunting in Europe is a conservation case, not just a moral one; it raises fundamental questions about what land is for, and in particular, what duties come with the privilege of land ownership.

Society has, in a sense, been repositioning nature in the cultural life of Britain for years, and, I discovered, in other countries too. Some years ago the *Oxford Junior Dictionary* deemed, based on a supposedly academic assessment of children's word use, that 'adder', 'buttercup' and 'conker' no longer warranted a place in the *Dictionary*, and they were removed to make room for 'attachment', 'broadband' and 'celebrity' instead. Out went 'acorn', 'blackberry' and 'catkin'. In came 'analogue', 'BlackBerry' and 'chatroom'. 'Ash', 'bluebell' and 'cowslip', and another ninety-odd words from nature, the countryside and farming, were culled.[6] I met with Oxford University Press, who rejected my pleas for the *Dictionary* to help rebuild nature literacy, rather than acquiesce to deepening nature illiteracy. The societal trends they were reflecting are alarming.

In 2009, Natural England published a report showing that fewer than 10 per cent of children regularly play in natural places such as woodlands, countryside and heaths, compared with 40 per cent three or four decades ago. We can be sure that such play as does happen is more closely supervised and less free, less exploratory. More didactic, less educational. As well as being a form of developmental deprivation, the lack of quality green environments is lowering society's horizons for the kind of relationship with nature that might be attainable in the future. Will today's children – those who still enjoy a connection to nature – look back from an even more impoverished future decade, dreaming of bringing back the richness of experience they remember from the 2010s, as I do the 1970s?

It chills me to think that in the decade or more that has passed since this first became a cause for concern, many of the ten- to fifteen-year-olds we were concerned about then have become parents of second-generation disconnected children. Meanwhile, a research review by King's College London[7] found that children who do spend time learning in natural environments perform better in reading, mathematics, science and social studies. Exploring the natural world 'makes other school subjects rich and relevant and gets apathetic students excited about learning'. Numerous studies in the UK and the USA have shown that spending time outdoors improves children's long-term memory, attention, self-control, self-awareness, behaviour, standards, motivation and personal development.[8,9]

Watching the flight of specks of thistledown is my earliest memory of wildlife. My mother told us they were fairies and I pretended to believe her. They, and in some ways we, were a product of war. We lived two or three floors up a 1960 Le Corbusier-influenced block of council maisonettes on London's Golden Lane Estate. Our garden was a narrow balcony on which my father grew tomatoes. Our playground was the broad courtyard below, its swings, climbing frame and the multiple-coloured lines of a netball-cum-five-a-side pitch. My older brother preferred to play nearby, at the place we called the bombsite, that part of Cripplegate that still comprised twenty-year-old piles of rubble and strafings of thistle seeds. Hitler's 1940 blitzkrieg had been phase one in the development of the Barbican Estate and its world-renowned arts centre; phase two started after we moved away.

A small group of black-headed gulls appeared in the air alongside our balcony one hard winter. My father threw bread and they caught it in flight, unerring, even when he made a sport of it to test their prowess. I saw in them a power and a delicacy, athletic and balletic, and the pearly tints of the London sky. I saw the infinite versions of themselves they shaped from each tail-twist and every stall and jerk of their wings. They left us, and the next day we waited for them to come, but they never did. I became a birdwatcher, and the local ducks grew fat on sliced white bread.

Our parents were an unlikely pairing of London Eastender and Fenland farm labourer. Unlikely, that is, were it not for war, and the Women's Land Army. War, by its aftermath, brought my parents together, introduced me to wildlife in Cripplegate, and in 1963, thanks to the Abercrombie plan to relieve housing pressure in the city, brought us to the thrush-haunted gardens of Swanley. It was hardly the archetypal leafy suburb, Swanley, but a railway town, the first town in Kent after you leave the London boroughs behind. When I was old enough to explore alone, I found myself two regular haunts within walking distance. Bourne Wood was alongside the railway in the direction of London. I came to know every tree, path, mound and pond. Towards the genteel village of Crockenhill was a peculiar duckweed-covered pond, amid a tangle of ancient willows that grew at all angles but the vertical. It was about 30 yards across, and so thickly vegetated that it could only be viewed from the lane that bordered it on one side, until I gradually cleared a twisting path that only I could see.

These places were the haunt of slow-worms and crested newts, abundant enough to be kept as pets by many of my school friends. Along with the white-clawed crayfish we found without difficulty in the River Darent at Eynsford, they are today highly protected species. Eynsford is a fair walk from Swanley, or one stop along the railway line that leads away from London and deeper into Kent. I would walk, sometimes alone, sometimes with my sister or with school friends, to the river where the alders attracted siskins in winter, then back to Eynsford station and the train home. I doubt if my parents knew where I was half the time.

After Doncaster, the London train runs alongside Potteric Carr, a wetland managed by the Yorkshire Wildlife Trust. Tufted ducks and coots dot the surface of Low Ellers Marsh but the train has, as always, picked up too much speed to allow any more exotic discovery. After another 8 miles the River Idle washlands flash by, wet-flushed pasture land with redshanks, voiceless against the train's din, and black cattle meeting around a hay clamp. The land flattens into the Trent floodplain, intensive arable country, and we cross the two arms of the River Trent, one natural, the other built to route sugar and wool from Newark to the Humber and the North Sea. The view from this train is a *vade mecum* to the English East Midlands that is too familiar to shake me from my brow-heavy torpor. Until, at the southern edge of urban Grantham, a red kite gyrates around a tight circle, close enough to create a graphic blazon against the window-framed sky. The bend of its wings, the deep fork of its tail, its hang, make of it an ideogram, like a flashed reminder. In view for a few seconds, it becomes a token of possible journeys to come.

When I was fourteen, my parents took us on holiday to Wales because I wanted to see a red kite. We spent a week looking for them, and finally, a distant view of a single bird over Tregaron Bog was all the reward I wanted. By centuries-long attrition we had lost a bird of peopled places and created a bird of remote wilds. But the red kite – more than any other threatened species – is now restored to its proper place, and into an enlightened society that celebrates its return. The red kite was a well-chosen subject for an early reintroduction programme, a true bird of the people, one that succeeds because of its tolerance of our imperfections. It symbolises success in these terms, while pointing to the toughness of the challenge ahead, if any wider relationship with nature is to be restored.

At Creeton, another kite, red as a fox, enters the airspace. A thought enters my head. I have been travelling through a contiguous, overlapping amalgam of places. Their names and histories are irrelevant to policy-makers for whom land is a careful tangle of criteria and parameters, boundaries and acreages. I remember Siôn Dafis's Ordnance Survey map of Snowdonia and its handwritten trove of unregistered titles. What would it cost to restore these names to the land? More than just a reprint: a repurpose. Billions?

Well, we have billions. The UK public purse pays out £2.4 billion* a year to a single land-based industry: farming; with the biggest landowners (or land investment companies) getting the largest share. The public benefit from this largesse is dubious: the government's Natural Capital Committee estimates that in England alone agriculture generates a net external cost to society of £700 million a year.[10] Of all the possible consequences of Brexit, the easiest to predict is the cutting of agricultural subsidies. But by redefining them instead, that £2.4 billion of tax money can be made to work harder for people, as an investment programme for the land

* 2015 Basic Payment Scheme figures, allocated on a per-hectare basis to each farm; this does not include payments for specific activities such as agri-environment measures.

assets of the nation, while still being available to boost incomes on those farms where wider public benefits are being delivered.

Conservation in the twenty-first century is about developing, or restoring, a healthy and mutually enriching relationship between people and nature, everywhere – in towns, cities, the countryside and the wild. But public policy insists on compartmentalising land, and compartmentalises land apart from people. Public policy lacks any sense of place, and has helped fuel people's detachment from place. It gives built-in advantage to the would-be destroyers of places they have never walked in, listened to, been rained on in, written poems about.

I am dismayed by the referendum result. But the truth is that while we continue to struggle to (merely) accommodate nature in modern life, new models are arising all the time in other developed countries. Should significant natural features have legal rights, as have been declared for the Whanganui River and Mount Taranaki in New Zealand and the Ganges and Yamuna in India? Should we protect our sonic environment by gazetting the most important natural and cultural soundscapes, as they have in Japan? Recent thinkers such as George Monbiot have made a passionate case[11] for a more untamed, 'rewilded' Britain, inspired in part by a renewed relationship with apex predators (reintroduced wolves) in Yellowstone National Park. When Britain (and most of Europe) last had wolves to contend with, it was our mediaeval beliefs and terrors that were the real enemy. In our more informed times, even on a small and crowded island, we may yet make room for them, if first we can make room for them in our heads.

Early in these journeys I discovered Maria Àngels Anglada, the Catalan novelist and poet who grew up speaking an illegal language. When the ice of repression receded, her words rebounded off the page, and the words she chose to use were those of nature. Olivier Messiaen, the composer whose blue rock thrush I sought and found on the coast near Banyuls, wrote his first significant birdsong-inspired piece in Stalag VIII-A prisoner-of-war camp in Görlitz (now Zgorzelec, Poland). *Quartet for the End of Time*, a staple of modern repertoire, was originally written for Messiaen and three fellow prisoners to perform, which they did on 15 January 1941 to a rapt audience of 400 inmates. For both Anglada and Messiaen, wildlife symbolised freedom and identity, and their works mined deep reserves of personal and cultural connectedness to nature. When Lönnrot was writing the poetry and Sibelius the music that would inspire the creation of independent Finland a hundred years ago, wildlife was an inextricable part of the narrative.

Our cultural severance from nature has coincided with its dramatic decline and the increasing ubiquity of isolating pursuits led by consumer technology. Nature has become commodified – as an attraction, or an education resource, or something children are taken to, to be supervised while they watch it or dip their nets for it. We have to believe (all conservationists are optimists) that this severance need not be total, nor irreversible. But we do need to re-imagine the business of conservation as one of several interdependent land-based industries, and put people back in the landscape.

The UK now has an opportunity to support land free of the arbitrary restrictions, idiocies and injustices of the Common Agricultural Policy (CAP). Rather than obscuring, institutionalising and rewarding bad land management, post-Brexit subsidies must invest in land as a national asset, invite and embrace public scrutiny and be held to account. Food production, wood production, water, carbon and wildlife can be re-amalgamated and supported in a more integrated way: the key is not to concern ourselves with attaching administrative labels to land and the people who occupy it. The agriculturally unproductive and dysfunctional uplands, and especially the National Parks, along with floodplains and coastal areas fighting hopeless battles with the sea, are obvious places to experiment with new models. If the UK can develop a subsidy system that is economically, politically and environmentally sustainable, it could even be a basis for the eventual replacement of the CAP, and therefore the reconnection of people and nature, across the EU and beyond.

Notes and Selected References

Spain

1. Abel Chapman, *Retrospect: Reminiscences and Impressions of a Hunter-Naturalist in Three Continents 1851–1928* (London: Gurney and Jackson, 1928).
2. Guy Mountfort, *Portrait of a Wilderness* (Newton Abbot: David & Charles, 1968; 1st edn Hutchinson & Co., 1958).
3. The following winter was one of the wettest in recent years, with nearly 12 inches of rain (300 litres per square meter) falling between September and December 2016.

4,5. Miguel de Cervantes Saavedra, *Segunda parte de El ingenioso hidalgo don Quixote de La Mancha*, 1615, chapter XII, (trans. Laurence Rose) [part of a list of useful things man has learnt from animals, distantly paraphrasing Pliny].

6. *No han destruït aquest recer vivent que ja lluny enyoren tantes ales. Aquí retroba l'aigua nodridora i els verds amagatalls l'ocell del nord* (trans. Laurence Rose).
7. Laia Castells & Paula Torramilans, *El niu dels noms: recull dels noms populars dels ocells a Catalunya*, student research project, Institut Lluís Vives, Barcelona, 2014 http://www.edubcn.cat/rcs_gene/treballs_recerca/2013-2014-03-2-TR.pdf.
8. *Recorda sempre aquesta claror / gràvida d'ocells quasi tots invisibles / fora d'una cardina que es gronxa / a la branca més alta del més alt xiprer.* Mª. Àngels Anglada, *Des de les Closes* (trans. Laurence Rose).

France

1. Olivier Messiaen (1908–1992): programme note written on the score of *Le Merle Bleu* (Blue Rock Thrush) for solo piano (trans. Laurence Rose).
2. Messiaen referred to herring gulls (*goéland argenté*), which was considered correct at the time; he would have seen the yellow-legged Mediterranean form, which is now known to be a distinct species (*goéland leucophée*).
3. A. Tamasier & P. Grillas, 'A review of habitat changes in the Camargue: An assessment of the effects of loss of biological diversity on the wintering waterfowl community', *Biol. Conserv.*, 70 (1994), 39–47.
4. R. Mathevet & F. Mesléar, 'The origins and functioning of the private wildfowling lease system in a major Mediterranean wetland: The Camargue (Rhône river delta, southern France)', *Land Use Policy*, 19:4 (2002), 277–86.
5. As written, this is technically ambiguous – it would be more correct to say it was a slightly *wide* minor third; in other words, somewhere between the minor and major interval. A minor third is an interval

four semitones apart, such as between E flat and C; a major third is wider by another semitone, for example E natural to C.

6. Konrad Leniowski & Ewa Węgrzyn, 'The carotenoid-based red cap of the Middle Spotted Woodpecker *Dendrocopos medius* reflects individual quality and territory size', *Ibis*, 155:4 (2013), 804–13.
7. Yang Lian, *A Wild Goose Speaks to Me*, 2006, http://yanglian.net/yanglian_en/essays/essays_01_07.html.

United Kingdom

1. J. E. Kelsall & P.W. Munn, *The Birds of Hampshire and the Isle of Wight* (London: Witherby, 1905).
2. Letter to Daines Barrington 7 August 1778, in Gilbert White, *The Natural History and Antiquities of Selborne*, 1789 (London: White, Cochrane and Co., 1789; 1813 edn).
3. Kelsall & Munn, *Birds of Hampshire*.
4. Thomas Hardy (1840–1928), 'Shelley's Skylark' (1913), after Percy Bysshe Shelley's (1792–1822) 'To a Skylark' (1820); and George Meredith's (1828–1909) 'The Lark Ascending' (1881), inspiration for Ralph Vaughan Williams's (1872–1958) 1914 rhapsody.
5. Casaubon (*Ephemerides*, 19 September 1611), writing his diary in Latin, translated 'God' and 'wit' literally – *Dei ingenium*.
6. William Hall, 'Memories of a Decoy', in *A Chain of Incidents Relating to the State of the Fens from Earliest Accounts to the Present Time* (printed by W. G. Whittingham, Lynn, 1812).
7. Dimock (1810–1876) was among the numerous correspondents who contributed information to William Yarrell (1784–1856) for the three-volume *A History of British Birds* (1843). Others included Carl Linnaeus and Thomas Pennant, the recipient of many of Gilbert White's letters.
8. Alan R. Thomas, *The Linguistic Geography of Wales* (Cardiff: University of Wales Press, 1973).
9. The linguistic divide between English-speaking and Welsh-speaking communities in south-west Wales has been famously sharp for centuries. The area is sometimes known as Little England Beyond Wales.
10. Roger Deakin, *Wildwood – A Journey Through Trees* (London: Hamish Hamilton, 2007).
11. P. S. Thompson, D. J. T. Douglas, D. G. Hoccom, J. Knott, S. Roos & J. D. Wilson, 'Environmental impacts of high-output driven shooting of Red Grouse *Lagopus lagopus scotica*', *Ibis*, 158 (2016), 446–52.
12. As long ago as 1988, studies showed that biodiversity and, for some species, abundance were lower in managed (for grouse) heather than in unmanaged heather – see M. B. Usher & D. B. A. Thompson (eds.), *Ecological Change in the Uplands* (Oxford: Blackwell Scientific Publications, 1988).
13. J. Arizaga, M. Willemoes, E. Unamuno, J. M. Unamuno & K. Thorup, 'Following year-round movements in Barn Swallows using geolocators: Could breeding pairs remain together during the winter?', *Bird Study*, 62:1 (2015), 141–45.

14. *The Breeding Bird Survey 2015.* BTO Research Report 687. British Trust for Ornithology, Thetford. Trends in England, Scotland and Northern Ireland are taken from sample sizes large enough for the trends to be assessed confidently. In Wales the population appears to be stable or slightly decreasing, based on a smaller sample size.
15. C. A. Morrison, R. A.Robinson, J. A. Clark & J. A. Gill, 'Spatial and temporal variation in population trends in a long-distance migratory bird', *Divers. Distrib.*, 16 (2010), 620–27.
16. C. A. Morrison, R. A. Robinson, J. A. Clark, A. D. Marca, J. Newton & J. A. Gill, 'Using stable isotopes to link breeding population trends to winter ecology in Willow Warblers, *Phylloscopus trochilus*', *Bird Study,* 60:2 (2013), 211–20.
17. C. M. Hewson, K. Thorup, J. W. Pearce-Higgins & P. W. Atkinson, 'Population decline is linked to migration route in the Common Cuckoo', *Nat. Commun.*, 7 (2016), 12296 doi: [10.1038/ncomms12296].
18. William Henry Hudson, *Rare, Vanishing and Lost British Birds* (London: J. M. Dent and Sons, 1923).
19. Henry Seebohm, *A History of British Birds, with Coloured Illustrations of Their Eggs* (London: R. H. Porter, 1883).
20. Later in 2016, in April, a goshawk was seen to be shot in the Strathdon area and was later euthanised. In June, two illegally set spring traps fatally injured a common gull on the Invercauld Estate. A spokesman for the estate issued a statement denying responsibility and implying that the evidence was planted in order to discredit what he called 'the grouse industry'. In August, newly fledged and tagged hen harrier 'Brian' disappeared near Kingussie. In March 2017, a tagged golden eagle disappeared in the Glenbuchet area. All these incidents were inside the Cairngorms National Park.

Sweden

1. J. Červený, S. Begall, P. Koubek, P. Nováková & H. Burda, 'Directional preference may enhance hunting accuracy in foraging foxes', *Royal Society Biology Letters* (2011), http://rsbl.royalsocietypublishing.org/content/7/3/355.
2. Within two months, another two golden eagles were added to this grim toll: one, tagged on 1 July 2014 at a nest in south Inverness-shire, was last recorded 4 June 2016; another, named 'Brodie' by local schoolchildren, was tagged on 26 June 2014 at a nest in east Inverness-shire. Her last recorded position placed her in the northern Monadhliath Mountains on 2 July 2016. In August, a recently fledged and tagged hen harrier named 'Elwood' disappeared on a grouse moor in the Monadhliaths. In June 2017, a buzzard was illegally trapped on Beinn Bhreac in the same range. Following these latest outrages, the Scottish Government commissioned a review of satellite-tagged eagles. The report, published in May 2017, showed that approximately one-third of tagged golden eagles fledging from Scottish nests are being illegally killed, with a clear link between these crimes and land intensively managed for driven grouse shooting – particularly in four

areas of the eastern and central Highlands. Recognising this, the Cabinet Secretary launched an independent enquiry into the environmental impact of grouse moor management and options for regulation.

Finland

1. Francis Willughby (1635–1672), *Ornithologiae libri tres* (Ornithology book three), completed by John Ray (1627–1705).
2. Greenpeace, *Certifying Extinction? An Assessment of the Revised Standards of the Finnish Forest Certification System,* 2004, http://www.sll.fi/mita-me-teemme/metsat/tiedostot/certifyingextinction.pdf.
3. Elias Lönnrot, *Kalevala*, Rune V, (1849, trans. John Martin Crawford, 1888).
4. The Arctic Circle is the line where the centre of the sun fails to dip below the horizon exactly once a year, but the size of the sun's disc, and atmospheric refraction, mean that this is a geometrical rather than observable phenomenon.

Norway

1. Nan Shepherd (1893–1981). *The Living Mountain*, written in the 1940s but not published until 1977.

Missed Connections

1. R.F.A. Grimmett & T. A. Jones, *Important Bird Areas in Europe*, International Council for Bird Preservation Technical Publication no. 9, (1989). The IBA concept has since been adopted worldwide – the latest global data set can be found at http://datazone.birdlife.org/site/search>. BirdLife International is a global partnership of national conservation NGOs, including the RSPB in the UK. It is represented in 120 countries as at the end of 2016. The original IBA 'orange book' was published by its predecessor, the International Council for Bird Preservation.
2. The most recent site condition information is given on the websites of Natural England, Scottish Natural Heritage, Natural Resources Wales and Department of Agriculture, Environment and Rural Affairs for Northern Ireland. However, the amount of information and detail seems to vary between them, and comparison between the four countries is not possible because the 95 per cent favourable condition target is based on differing criteria (area in England, proportion of sites in Wales, proportion of designated features in Scotland) and timescales.
3. This refers to burning of heather as part of a grouse management regime. As grouse moor management has become more intensive, burn rotations have become shorter. It is a practice that is widely criticised for its impact on peat soils, vegetation, water quality and, in intensively managed areas, flood risk. See P. S. Thompson, D. J. T. Douglas, D. G. Hoccom, J. Knott, S. Roos & J. D. Wilson, 'Environmental impacts of high-output driven shooting of Red Grouse *Lagopus lagopus scotica*', *Ibis*, 158 (2016), 446–52.

4. The State of Nature partnership, *State of Nature 2016*, https://www.rspb.org.uk/Images/State%20of%20Nature%20UK%20report_%2020%20Sept_tcm9-424984.pdf.
5. Royal Society for the Protection of Birds (RSPB), 2016. Written evidence to UK Parliament Petitions Committee in advance of a Westminster Hall debate on driven grouse shooting: http://data.parliament.uk/writtenevidence/committeeevidence.svc/evidencedocument/petitions-committee/grouse-shooting/written/40095.html.
   Natural England data supplied to the RSPB under Environmental Information Regulation request 2744 (February 2015) indicated that over a ten-year period more than £105 million of agri-environmental funding supported management systems that carry out burning of blanket bog habitat on grouse moors. Almost all Sites of Special Scientific Interest (SSSI) units with consent to burn blanket bog in the Special Areas of Conservation (SAC) designated for this habitat in England were consented under Higher Level Stewardship agri-environmental funding agreements.
6. I have blogged extensively on the *Oxford Junior Dictionary*. Details of the word changes can be found at http://www.naturemusicpoetry.com/uploads/2/9/3/8/29384149/words_taken_out.pdf.
7. Kings College, London, *Understanding the Diverse Benefits of Learning in Natural Environments*, (Commissioned by Natural England, London, April 2011), http://www.naturalengland.org.uk/Images/KCL-LINE-benefits_tcm6-31078.pdf.
8. Stuart Nundy, *Raising Achievement Through the Environment: The Case for Fieldwork and Field Centres* (National Association of Field Studies Officers, 2001, cited by Kings College, London, as above), http://bit.ly/1fdazsx.
9. William Bird, *Natural Thinking: Investigating the Links between the Natural Environment, Biodiversity and Mental Health* (Royal Society for the Protection of Birds, 2007), http://www.rspb.org.uk/images/naturalthinking_tcm9-161856.pdf.
10. The Natural Capital Committee's report *The State of Natural Capital: Protecting and Improving Natural Capital for Prosperity and Wellbeing* (2015) states: 'Farming can produce large external costs to society in the form of greenhouse gas emissions, water pollution, air pollution, habitat destruction, soil erosion and flooding. These costs are not reflected in the price of food. As a result, farming is responsible for net external costs to society that have been valued at £700m per annum.' The valuation was based on the government's Environmental Accounts for Agriculture work which was discontinued in 2010. Net external cost figures were last produced for 2008.
11. George Monbiot, *Feral* (London: Allen Lane, 2013).

Comprehensive scientific and other references can be found at www.thelongspring.com/the-book

Acknowledgements

I have worked for the RSPB since 1983, and during 2016 I worked on various short-term, part-time projects, which gave me the flexibility to spend some of the year travelling for this book, and time to write it. I am hugely grateful to the RSPB for allowing me this flexibility.

Some of the plants mentioned were only identified after I returned to the UK with the help of my RSPB colleague Tim Melling, who also helped with a wealth of other facts and information. Two other colleagues, Helen Byron and Ian Thompson, as well as Alison Duncan of the Ligue Française pour La Protection des Oiseaux and Gemma Rogers of the Campaign for National Parks, also generously gave of their time, discussing various conservation issues that I wanted to address. The views expressed are mine, and any errors mine, too.

I was delighted to come across the poetry and prose of Maria Àngels Anglada, and hope to play a small role in bringing her to the attention of English-language readers. Her daughters Mariona and Rosa Geli Anglada kindly gave me permission to quote and translate her work; and most importantly, along with their aunt – the poet's sister Enriqueta Anglada d'Abadal – commented on and improved my efforts. Another RSPB colleague, Carles Carboneras, helped greatly, too. Professor Mariàngela Vilallonga of the University of Girona (Càtedra de Patrimoni Literari M. Àngels Anglada – Carles Fages de Climent) put me in touch with the Anglada family, and also with Carles Fages Torrents, who gave permission for me to quote and translate lines by his grandfather Carles Fages de Climent. Jordi Canet Avilés of the Ajuntament de Castelló d'Empúries gave helpful guidance. John Starr kindly gave permission to quote from the unpublished diary of his relative Dr Katherine Heanley.

Joan Childs, Ulla Matturi, Kirsi Peck and Ian Rotherham helped in various ways, with apologies to anyone I have omitted. It has been a pleasure working with the artist Richard Allen in what has proved to be a genuine collaboration, and with Julie Bailey of Bloomsbury. I am grateful for Julie's patient help to a novice author and for being so open to my ideas. Bloomsbury's designer, Jasmine Parker has been hugely helpful in shaping the overall look of the book.

During my travels I met many helpful people, but I particularly want to thank José Navarrete Pérez (Spain), Siôn Dafis (Wales), Håkan Vargas (Sweden), Minna Pyykkö (Finland) and the late Ralph Sargeant (Fenland) for their time and excellent company. Jane Rose was delightful company in Doñana, the South Downs and the Scottish Highlands. Jane commented comprehensively on my first draft and tolerated my frequent absences – and, for that matter, my presences.

Index